327.51047
W63s

118504

DATE DUE			
FEB 14 1985			

Sino-Soviet Crisis Politics
A Study of Political Change and Communication

HARVARD EAST ASIAN MONOGRAPHS
96

SINO-SOVIET CRISIS POLITICS
A Study of Political Change and Communication

RICHARD WICH

CARL A. RUDISILL LIBRARY
LENOIR RHYNE COLLEGE

Published by COUNCIL ON EAST ASIAN STUDIES, HARVARD UNIVERSITY and distributed by HARVARD UNIVERSITY PRESS, Cambridge (Massachusetts) and London   1980

© Copyright 1980 by the President and Fellows of Harvard College

The Council on East Asian Studies at Harvard University publishes a monograph series and, through the Fairbank Center for East Asian Research, administers research projects designed to further scholarly understanding of China, Japan, Korea, Vietnam, Inner Asia, and adjacent areas.

Library of Congress Cataloging in Publication Data

Wich, Richard, 1933-
Sino-Soviet crisis politics.
(Harvard East Asian monographs ; 96)
Bibliography: p.
Includes index.
1. China—Foreign relations—Russia. 2. Russia—Foreign relations—China. I. Title. II. Series.
DS740.5.R8W53   327.51047   80-15409
ISBN 0-674-80935-1

To the Memory of My Father

Contents

ACKNOWLEDGMENTS ix

1 INTRODUCTION 1
2 POLITICS OF THE BORDER QUESTION 25
3 IMPACT OF CZECHOSLOVAKIA, 1968 41
4 EAST EUROPE AND SINO-SOVIET BORDER TENSION 65
5 CHINA, VIETNAM, AND THE UNITED STATES 75
6 THE BORDER CLASHES 97
7 THE NINTH CCP CONGRESS 113
8 THE MOSCOW CONFERENCE 123
9 THE SUMMER OF SIXTY-NINE 163
10 INTERNATIONAL COMMUNIST POLITICS 193
11 ONTO THE NEGOTIATING TRACK 207
12 TRANSFORMATION OF ADVERSARY RELATIONS 231
13 CONCLUDING PROPOSITIONS 269

ABBREVIATIONS	287
NOTES	288
BIBLIOGRAPHY	307
INDEX	309

Acknowledgments

Through an analysis of political perception and communication, this study seeks to account for the structural change in the international political landscape that followed the Soviet intervention in Czechoslovakia in 1968. It is central to this account that the meaning of Czechoslovakia, in its fundamental effects, must be understood in a broad context of international and (on the Chinese side) internal politics. Accordingly, such a study must draw on a wide range of sources in order to follow the complex interplay of perceptions and reactions that were communicated by the parties involved. Those sources are accessible in a real sense because of the informed labors of numberless monitors, linguists, and editors who have compiled documentary records that enable a researcher to find his way through a massive amount of data. I am pleased to record my gratitude for their efforts.

Happily, I can name those other analysts who have also dealt with these sources and, in so doing, have made them manageable and meaningful for my own work. These include Arthur Berger, Avis Boutell, Joyce Forrest, Werner Hahn, Robert Sutter, and Frederick Tibbetts. Special acknowledgment is due Gwendolyn Baptist for the standards she has long represented of the kind of close reading and analysis that this work has sought to demonstrate. I am grateful to Ellis Joffe for good advice in determining the shape of this study, which was part of a larger undertaking that would have been too unwieldy in its original form.

At Harvard University, the East Asian Research Center provided a hospitable home, and the Council on East Asian Studies provided effective editorial support in the preparation of this work.

Sino-Soviet Crisis Politics
A Study of Political Change and Communication

ONE

Introduction

This is a study of the politics of crisis in the root sense of a turning point, a decisive stage in development whose outcome registered a geologic shift in the international political environment. This structural change involved relations among the most powerful centers of geopolitical gravity, the United States, the Soviet Union, and China. But, while this concerns a fundamental change, it must be stressed at the outset that a highly political process was at work, one of interacting perceptions and communications rather than the ineluctable unfolding of "deep," irreversible historical forces. Accordingly, this study very much concerns perception and communication, particularly the process of political "signaling" through which change is initiated and perceived.

The focus of the study is on the Sino-Soviet crisis that reached an acute stage at the end of the 1960s. That crisis is viewed here as symptomatic of a basic change, the surface eruption of underlying structural shifts. This study seeks to elucidate the meaning of the Sino-Soviet border crisis in terms of the shift in alliance and adversary relations distinguishing the global political environment emerging in the 1970s from the earlier post-World War II era. That shift may be viewed as successive to an earlier one, in which the Sino-Soviet alliance system was eroded and then collapsed. The 20th Congress of the Soviet Communist Party gave a sharp impetus to that earlier turning point, marked by the crisis of destalinization, which led to a process that shattered the Sino-Soviet axis of "the socialist camp." The starting point of the next stage, the point of departure for this study, was the Soviet-led invasion of Czechoslovakia. The interacting perceptions of that event produced the *politics* of the border crisis, from which emerged a shift in adversary relations no less significant than the earlier transformation of alliance relations.

It might be useful here to enumerate the effects of the structural change of which the Sino-Soviet crisis was an acute symptom. Some of these effects figure only rather tangentially in this study, and in any case their evolution required many years (and this is an open-ended process) beyond the critical phase treated in detail here. Nonetheless, these reflect the irradiating force of a systemic change in the global political environment. Analytically and somewhat artificially separated, these changes include the following, in descending order of direct relation to the crisis under study:

Soviet Union as China's primary adversary. The Sino-Soviet axis, which became severely strained in the late 1950s because of the two partners' divergent expectations from their alliance, collapsed from the pressures of the Vietnam conflict, the conclusive example of what the Maoist leadership in Peking perceived as Soviet readiness to subordinate other Communists' interests to Moscow's paramount concern over its relationship with the United States. After Czechoslovakia, in a major policy shift portending a new turning point, the Chinese undertook initiatives that served to identify the Soviet Union as China's primary adversary. At its most dangerous and dramatic, this took the form of the border crisis in 1969, a confrontation demonstrating that the Sino-Soviet rivalry had reached the most sensitive and explosive stage.

Sino-American accommodation. Immediately after Czechoslovakia, a parallelism of interest between China and the United States vis-à-vis the Soviet Union was discernible, and a few months later, in one of a series of signals in a complex communication process between Peking and Washington, the Chinese made an exploratory overture to the incoming Nixon Administration. This process was not only a corollary but an extension of the deepening Sino-Soviet confrontation. Conditions were propitious in the United

States for a new Sino-American relationship, but this process was problematic and perilous for the Chinese leadership. Powerful elements in that leadership had strong vested interests in a dual confrontationist policy in which the Soviet Union shared with the United States the status of primary adversary. Those interests were overridden, though not without convulsive internal struggle, as the ascendant wing of the Chinese leadership followed through on the logic of the situation as perceived after Czechoslovakia.

Sino-Japanese reconciliation. The underlying structural changes involving the main geopolitical forces imparted a powerful impact on Sino-Japanese relations, which underwent a sudden turnaround. Within a mere few months after President Nixon's China visit, the Japanese had normalized relations with Peking.

Sino-Vietnamese divergence. With the internationalization of the Vietnam conflict, Peking's relations with Hanoi, as in the case of Hanoi's relations with Moscow, suffered some strain from competing priorities, but after Czechoslovakia a process emerged in which Sino-Vietnamese relations became increasingly estranged while Soviet-Vietnamese ties became tighter. This process was directly rooted in the pattern of events following Czechoslovakia; indeed, it was through the Vietnam prism that the meaning of Czechoslovakia acquired its sharpest focus for the Chinese, so that the deterioration of Sino-Vietnamese relations was an integral part of the geologic changes underway in the late 1960s. (It will not do to explain Sino-Vietnamese conflict as simply a natural product of historical forces. Before the turnaround in the Sino-Japanese relationship, that too had been viewed as a natural rivalry with a long future as well as past.)

China's close relations with Romania, Yugoslavia. One of Peking's early initiatives in the wake of Czechoslovakia was to fashion an active, flexible East Europe policy centered on the Balkans, a region where resistance to Soviet dominance was strongest. China's Albanian connection had been present for years, but now the Sino-Romanian relationship flowered into a close friendship and—most notably and revealingly—the venomous Sino-Yugoslav quarrel was transformed overnight. The change proceeded through a diplomatic phase, then a more broadly political one, and finally even acquired an ideological cast as the Chinese began to investigate the Yugoslav internal system as having features worth emulation.

Sino-Albanian estrangement. In its early phase, Peking's refurbished East Europe policy laid stress on Sino-Albanian mutual security interests. Had the Chinese persisted in their dual confrontationist policy, this mutuality could have survived, but it fell victim to the logic of the situation that propelled China and the United States into a new relationship. In this changed context, the top Albanian leaders perceived Peking's Balkan policy as inimical to their own interests, and they sharply dissented from Peking's move away from a dual confrontationist policy. The Sino-Albanian relationship had thus moved onto a slippery slope leading to estrangement.

Cuba as a close Soviet ally. Of all those enumerated, this development was the most remote from Czechoslovakia as the starting point of structural change. The failure of Havana's insurrectionary policies in Latin America, as dramatized by Che Guevara's death in Bolivia, reduced the Cuban challenge to Moscow's line, and to Moscow's client parties in the region, that had strained Soviet-Cuban relations. Nonetheless, Fidel Castro's reaction to the Czechoslovak intervention, no less interesting than Hanoi's and Peking's, clearly registered his perception of the broader significance of that event. As such, his reaction signaled a turning point in Soviet-Cuban relations, which evolved in such a way that in the following decade the Cubans could be widely perceived as Soviet proxies in the broader geopolitical struggle.

THE POINT OF DEPARTURE

As the above effects illustrate, the invasion of Czechoslovakia—or more precisely, the perceived implications of that event—provide the context for this study's point of departure. It was in this context that Peking began to evolve new policy lines which converged in the border issue that soon surfaced and developed into a severe crisis in Sino-Soviet relations. One was a new, differentiated, and flexible East Europe policy, dramatically illustrated by a rapid movement toward good relations with that erstwhile arch-heretic, Yugoslavia's Tito. The Chinese were concerned over any propensity Moscow might have had after the crude crushing of Czechoslovakia to proceed to put all of East Europe in order, which would have meant squelching Romanian and Yugoslav independence and bringing to heel Peking's ally Albania. Peking thus sought to promote whatever countervailing forces in East Europe it could find, departing from its previous approach of rejecting or, at best, remaining on cool terms with East European regimes that did not show proper sympathy for the Chinese ideological line in the Sino-Soviet rivalry. Another new Chinese line was a refurbished East Asia policy, one that was also marked by a significant flexibility, in sharp contrast to the rigid stance Peking had assumed in the course of the Vietnam conflict. One key element of this new policy was an openness to a negotiated settlement in Vietnam, a departure from Peking's theretofore adamant insistence on a no-compromise military resolution of that protracted conflict. A second, closely related element was an overture to the incoming Nixon Administration to explore possibilities of a new relationship, replacing the frozen hostility between the United States and the People's Republic of China. These moves, complemented by United States troop withdrawals from Vietnam and American initiatives toward an improved Sino-U.S. atmosphere, were designed to prepare the ground for new relationships in East Asia following a Vietnam settlement.

As this study will show, the Sino-Soviet border crisis originated in the context of Peking's new policy directions arising from the interacting perceptions of the invasion of Czechoslovakia. As part of the effort to deter further Soviet moves in East Europe and to put teeth into Chinese declarations of support for the vulnerable East Europeans, Peking developed a campaign linking portrayals of a Soviet threat in East Europe with warnings of Soviet troop concentrations and provocations along the Chinese border. Using aggressive patrolling of disputed areas to underline border tensions, the Chinese were able to deploy a concrete deterrent against Soviet aggression in East Europe: a reminder that the Chinese could make trouble along a long border for which, as the Soviets had been sharply reminded earlier in the decade, a final territorial settlement was still far from having been achieved.

The new East Asia policy, looking toward a post-Vietnam War environment, was motivated in large measure by Peking's acute concern over any opportunity for the Soviets to strengthen their influence in that region. Two reinforcing perceptions were to play on this concern: one, of a declining United States presence with the capping of the American troop strength and signs of American war fatigue (President Johnson having wearily forsworn running for re-election), and the other, of a danger that the USSR would seek to fill the vacuum left by an American withdrawal. The aftermath of the invasion of Czechoslovakia had an important impact on the evolution of Peking's approach. In particular, Peking was infuriated and dismayed by the strong endorsement of the invasion given by the North Vietnamese, all the more so in that Hanoi coupled Soviet intervention in Czechoslovakia with support for the North Vietnamese as parallel efforts to preserve the integrity and security of "the socialist camp." In the years of escalating hostilities in Vietnam since the middle of the decade, Peking, at great cost diplomatically and politically in the Communist movement, had intransigently resisted all pressures (including strong internal ones) to reduce Sino-Soviet antagonism in the interest of socialist camp solidarity with the embattled Vietnamese comrades. Now the Chinese perceived Hanoi's reaction to Czechoslovakia as a warning signal of an opportunity for Moscow to consolidate its political credit from its arms-supply relationship with the Vietnamese and to enhance its position in Indochina as the war wound down. Having incurred such a high cost in spurning socialist camp solidarity in the preceding years, despite intense pressure from within the Communist movement, the Chinese were in no mood now to watch indifferently while Moscow consolidated its gains and entrenched itself in the new circumstances.

Consequently, Peking was intent on making every effort to scuttle Moscow's opportunities and to serve notice that the new challenge would be met by a new response. Peking's response followed a rapidly evolving course. Less than two weeks after the invasion of Czechoslovakia, the Chinese reacted with polemical fury to Hanoi's endorsement of the Soviet intervention; at the same time, the Chinese were developing a new phase in their anti-Soviet campaign, one that was keynoted by attacks on Soviet "social-imperialism," a term designed to remind the world that the current Kremlin leadership was the successor to a Russian imperial expansionism of which China was but one of the victims. Less than a month after Czechoslovakia, Peking issued an ominous signal by raising the issue of border tensions, a development that not only served notice on the Soviets that tensions on a second front would make further aggression in East Europe risky, but was also designed to provoke the Soviets into demonstrating their "social-imperialist" nature in Asia. At the very least, Peking's linkage at this time of Soviet behavior in East Europe with border tensions in Asia provided a sharp counterpoint to Hanoi's coupling of the intervention in Czechoslovakia with support for the Vietnamese Communists.

Thus, Peking had decided in the wake of Czechoslovakia to mount a forceful response to what it perceived as a challenge of Soviet expansionism ("social-imperialism"). This was a different sort of response from the ideological rivalry earlier in the decade when Peking challenged Moscow's leadership of the Communist movement; and it was diametrically opposite to the isolationist withdrawal in the middle of the decade when Peking's relations with most of the world fell into disrepair and ill repute. Now Peking was taking a flexibly assertive position politically and diplomatically, while subordinating the ideological considerations that had been paramount formerly, as signaled by the abrupt turnabout toward good relations with Tito (who had symbolized treachery a decade earlier when Peking was propounding the necessity of socialist camp discipline). This new approach was highly differentiated and varied but, in choosing at this time to make an issue of the border question, the Chinese surely selected the most direct possible challenge to Soviet expansionism. This challenge became pivotal to the structural change in the international system that had derived its initial impetus from interacting perceptions of the implications of Czechoslovakia.

THE CONCEPTUAL CONTEXT

If perceptions of the meaning of Czechoslovakia served as the point of departure for the events leading to the Sino-Soviet border crisis, these perceptions were shaped by conceptual frames structured in the international political environment emerging after World War II.[1] The dominant feature of that environment was the overarching bipolar global line of confrontation dividing the superpowers, the United States and the Soviet Union. Berlin, Korea, Taiwan, Vietnam—a roll call of focal points of international tension since World War II brings to mind the broader strategic meanings of these conflicts as perceived by the major powers. In the cases cited, the global line of bipolar confrontation divided countries or halted revolutionary movements short of their territorial fulfillment, thereby causing local conflicts to become polarized within the global power struggle. As a result, these became focal points of global conflict by virtue of their implications for the integrity and credibility of great-power commitments; by the same token, they became focal points for the testing of the internal alliance relationships in which they were enmeshed.

These paradigm cases, representing divided countries whose lines of division were stress lines in the bipolar system, were lumped in a single category that, in fact, was basically ambiguous, an ambiguity that much of a quarter century of postwar history was devoted to resolving. This would seem to be an instance of what logicians call a category mistake; in sorting out the ambiguity, a critical analysis would look beyond the fact that these were cases of divided countries (Taiwan being regarded by both sides of the Chinese Civil War as a part of China) and separate them into two categories. One category would include Germany and Korea, where the division was the direct product of the stationing of the Western and Soviet armies at the conclusion of the war. China and Vietnam would fall in the other category, as countries where

Introduction

revolutionary civil war became arrested by supervening forces, and the perception developed that issues in the global struggle were involved that transcended the local stakes. Rather ironically, this perception, and the assimilation of the China and Vietnam questions with the other divided countries, had as background a fateful ambiguity surrounding the great powers' stakes and commitments in Korea, the United States having left it unclear (at least) whether the writ of its commitment in that region ran as far as Korea or stopped with Japan. The outbreak of war on the Korean peninsula resolved that ambiguity by clearly drawing the division of Korea into the bipolar line of global conflict. But the perceived implications of the outbreak of war in Korea went far beyond that peninsula. The war also drew China deeply into the bipolar structure in two ways: by the United States decision, shortly after the outbreak of hostilities, to intervene in the Chinese Civil War by an interdiction of the Taiwan Straits (the Truman order of 27 June 1950 thereafter being marked by Peking as the beginning of the U.S. "occupation" of Taiwan), and the Chinese intervention four months later to preserve the division of Korea after General MacArthur's forces had swept northward to the Chinese border. Not only had Korea and the Taiwan question become deeply embedded in the bipolar structure, but it should also be noted that Truman's 27 June order included measures to enhance American support for the French against the Vietnamese Communists, a significant step in what would become a series of steps leading to full-scale United States involvement in the Vietnam War on the premise that issues having global implications were involved.

That the Taiwan question had been drawn deeply into the bipolar gravity had obviously important effects on Sino-American relations (which became more so when the United States signed a defense treaty with the Nationalist Chinese regime in 1954), but it would also have decisive implications for the Sino-Soviet relationship, a relationship that, with the formation of the Sino-Soviet alliance in February 1950, had constituted the axis of "the socialist camp" in the two-camp bipolar world view. Two fundamental, interdependent interests of the new Chinese Communist regime—

security of its borders and completion of the revolution—had in the first year of the regime's existence become subject to powerful forces of bipolarity by virtue of the Sino-Soviet alliance (embodying Mao's famous decision to "lean to one side" in the Cold War) and the outbreak of the Korean War. Increasingly, the Sino-Soviet relationship would be strained, and eventually fractured, by systemic forces which ideological, strategic, economic, and other bonds of alliance could not withstand. (The constraints and implications of bipolarity provide the conceptual context for the divergent expectations from the alliance system that produced the Sino-Soviet schism and eventually led to the transformation of alliance and adversary relations of which the border crisis was an acute symptom.) In this context, the geopolitical weight of an issue drawn into the bipolar gravitational system far transcended its intrinsic meaning, as, for example, the meaning of the Taiwan issue for Sino-American relations; perceptions of such issues were shaped by their place in the conceptual structure of alliance and adversary relationships formed by the bipolar forces dominating the international environment following World War II.

There was a period following the end of the Korean fighting when bipolarity was considerably attenuated. The Geneva accords on Indochina, the Bandung Conference and its celebration of peaceful coexistence, the Geneva Summit Conference, the Soviet-Yugoslav rapprochement, the opening of Sino-U.S. ambassadorial talks and Peking's line of "peaceful liberation" of Taiwan—these developments in the mid-1950s marked an era of relative détente and recourse to negotiation and conciliation across and beyond bloc lines. There were contrary developments that Peking, in particular, might have perceived as more in the Cold War tradition: the United States treaty with Nationalist China, Washington's distaste for the 1954 accords on Vietnam, the formation of the Southeast Asia Treaty Organization. These developments were undoubtedly influential in Peking's decision to take more forceful measures on the Taiwan issue and, more generally, to begin to insist on a sharply bipolar world view. The Sino-Soviet relationship itself, however, was more immediately affected by the structural tremors reverberating through the Communist system from Khrushchev's destalinization campaign launched at the Soviet 20th Party Congress in early 1956 and the Hungarian revolt later that year, the most explosive effect of those tremors. Alarmed by the Kremlin's clumsy handling of the matter, Peking began to play a mediatory role in Soviet bloc affairs, thereby introducing strains in the Sino-Soviet relationship that would be multiplied when, during the following year, Peking took a markedly tougher centralist line in bloc affairs coupled with a correlative emphasis on the bipolar nature of the international system.

During this watershed period, in the late 1950s, the Sino-Soviet relationship became a line of stress—and in the 1960s a fault line—in the international system, especially in respect to the integrative functions of the system. The bipolar system as a whole, in order to avoid nuclear catastrophe to the entire system, had become subject to various regulatory mechanisms (Western nonintervention in the Hungarian crisis, for example) and to efforts to explore patterns and norms of coexistence. Peking's concern to protect its interests from the effects of destalinization had prompted its active mediation in bloc affairs, but more fundamental strains in the Moscow-Peking axis began to emerge as the divergent perceptions of the operation of the international system became sharply focused within the bipolar framework. For Peking, the correlative stress on bipolarity and bloc centralism was designed to bolster socialist camp solidarity and support for Chinese security while a more aggressive, forward strategy was pursued against the United States, principally directed at the U.S. denial (dating from the Truman interdiction of the Taiwan Straits and formalized by the 1954 Mutual Security Treaty between Washington and Taipei) of Peking's claim to Taiwan. Peking saw the time as propitious for such a strategy, the Soviets having achieved significant technological breakthroughs in 1957 with the launching of the first ICBMs and sputniks.

For Moscow, however, bipolarity had quite different implications, Soviet strategic interests being mainly engaged in the relationships of competition and coexistence with the United States to which Chinese interests were subordinate. While Peking was looking to the bipolar system for support for a forward strategy against the United States, Moscow was exploring with Washington ways and means of stabilizing the system. Thus, on the critical (for the Chinese) issue of Taiwan, Peking perceived the operation of the system as sustaining a two-Chinas situation by imposing restraint in view of the two superpowers' commitments to their respective Chinese allies. System stabilization implied maintenance of the status quo. This was painfully evident to Peking from the Taiwan and Middle East crises of mid-1958, when the Soviets placed the interests of bipolar stability over the interests of anti-Western forces. Moscow's neutral stance on the Sino-Indian border dispute the next year, together with Khrushchev's pilgrimage to the United States to promote the détente "spirit of Camp David" with President Eisenhower, left little room for Peking to hope that its interests would be served by the existing system. And, lest the Chinese somehow fail to get the message, Khrushchev brought it to them directly from his visit to the United States when he warned his hosts in Peking of a lack of support for attempts to "test by force the stability of the capitalist system." Moscow also took care to limit Peking's ability independently to disturb the international system, notably by reneging on a nuclear-sharing agreement with the Chinese and by seeking to exercise control over Chinese military initiatives. Moscow later disclosed that Chinese shelling of the offshore islands in 1958 had been undertaken without consultation with the USSR, as required by the Sino-Soviet Treaty, and that Peking intentionally became embroiled in a border clash with India in 1959 to torpedo the pursuit of détente then taking place. Significantly, Moscow claimed that Chinese probes during this period were undertaken in areas—such as the Taiwan Straits and the Sino-Indian border—that could be manipulated by Peking in an effort to assume control over the strategy and policy of the entire socialist camp. As for the sensitive issue of nuclear weapons, Moscow would contend that "the very idea of the need to provide themselves with nuclear weapons could occur to the leaders of a country whose security is guaranteed by the entire might of the socialist camp only if they have developed some kind of special aims or interests that the socialist camp cannot support by its military forces."[2] Precisely to the point, which was that the interests of the leader of the socialist camp—a status Peking had insisted that Moscow assume—and those of its major ally were basically divergent and, thus, could not be accommodated by a common strategy. But, if Peking's interests would not be supported by the socialist camp, these interests could not be integrated into the bipolar system, and Peking would perforce need to undertake initiatives having destabilizing effects on that system. In the ensuing turbulent decade, these effects would transform basic alliance relationships and leave Peking groping for a way to cope with the constraints of bipolarity that now increasingly appeared inimical to its interests.

By the end of the 1950s, then, there was a sharp divergence in Moscow's and Peking's perceptions of their basic interests, a serious failure of expectations from the Sino-Soviet alliance, which would thereafter be subjected to severe strains the bipolar system could not contain. The Chinese thereupon proceeded to undertake, persistently and provocatively, a seemingly endless series of destabilizing initiatives that caused Moscow increasingly to share Washington's perception of Peking as an international outlaw. Having found that Moscow insisted on using its leadership of the socialist camp for its own interests, Peking mounted a vigorous ideological and political challenge to Soviet hegemony in the Communist movement, a challenge that, in effect, divided the Communist camp along its own East-West axis. Where the Chinese had, in the late 1950s, stressed the bipolarity of the international system as a whole, they now took Moscow to task for focusing solely on the "contradiction" between the two camps to the exclusion of other "fundamental contradictions" of the era. Significantly, the Chinese now insisted that the various types of contradictions were "concentrated" in Asia, Africa, and Latin America and that the "whole" revolutionary cause "hinges on the outcome of the revolutionary struggles of the peoples of these areas." While shifting the strategic focus away from the bipolar relationship as the decisive dimension of international affairs, Peking also challenged Moscow's effort to regulate and stabilize the international system through détente with the West. This challenge centered on the doctrine of peaceful coexistence, which, the Chinese bitterly charged, the Soviets had improperly elevated to a major premise of policy, thereby subordinating the unsatisfied interests of the Chinese, Korean, Vietnamese, and other Communist revolutionaries to the interest of system stability. It thus became Peking's strategy in the 1960s to resist the integrative pressures of the international system while assuming leadership of those forces having strong interests in promoting system destabilization.

It is in this context that the impact of the internationalization of the Vietnam conflict on the Sino-Soviet relationship can be understood. That impact confounded the conventional wisdom, according to which an external threat to an alliance system should strengthen the bonds of alliance. Indeed, that wisdom underlay the post-Khrushchev Kremlin leadership's line of Communist "united action" in support of the Vietnamese comrades under siege. Peking, however, saw Vietnam developments in a different conceptual context; its perception was expressed in the line that Moscow was seeking to bring the Vietnam conflict into "the orbit of U.S.-Soviet collaboration," which is to say that the Soviets were seeking to regulate the conflict in the larger interests of bipolar system stability. This accounts for the utterly rabid manner in which the Chinese opposed any initiative toward a Vietnam settlement, even initiatives supported by Hanoi itself. The impact of the powerful pressures generated by the Vietnam conflict caused the complete collapse of the Sino-Soviet alliance, with the Chinese going to extraordinary lengths to see to it that the residual bonds of alliance relations with Moscow were severed and that only adversary relations, with both the Soviet Union and the United States, remained in China's relationship to the bipolar system.

Thus, the latter half of the 1960s saw the completion of the process of alliance transformation that had its roots in the strains arising from the Soviet 20th Congress. The Sino-Soviet relationship was no longer a strained alliance or even a severely eroded one; it was now an adversary relationship on a par with the Sino-U.S. relationship (or, as Peking put it in referring to the United States and the Soviet Union, "the two No. 1 enemies"). Here enters Czechoslovakia, which, while paradoxically less severe in its direct repercussions than Hungary in 1956, proved to be the catalyst of a profound process of adversary transformation. Again, bipolarity provided the conceptual context, with Peking perceiving a systemic symmetry between Soviet intervention in Czechoslovakia (along with the Brezhnev doctrine of limited sovereignty generalizing that behavior) and the postwar United States interventionism that, from the time of the Chinese Civil War and the Korean War, and extending in recent years to large-scale involvement in Vietnam, had been regarded as fundamentally inimical to Chinese Communist interests. This perception, however, was now framed by a quite different East Asian situation from the set of circumstances obtaining in the middle of the decade. Whereas, at that earlier turning point, in 1965, the United States was beginning its direct combat engagement in the Vietnam conflict, by 1968 that involvement had peaked, and signs of disengagement were becoming evident. It was in this broader setting that Hanoi's strong endorsement of the Soviet intervention, particularly the conceptual context of that endorsement, had such a critical impact on Peking's perception of the meaning of Czechoslovakia. It should be stressed that the most significant dimension of Peking's reaction to Czechoslovakia was of a second order: it was reaction to others' reactions, principally the Vietnamese Communists, that had a transforming effect on Peking's conceptual and strategic thinking. That it was a second-order reaction reflected the systemic or structural implications of Czechoslovakia shaping Peking's response. Again, as in the 1965 turning point, the concept of a socialist camp (or "socialist community," Moscow's preferred term) within a bipolar framework was crucial; but now a process emerged in which Peking groped for a key to breaching the bipolarity by exploring ways of reintegrating itself into the international system (in contrast to its isolationist withdrawal in the previous phase) within a transformed structure of adversary relationships, principally China's reordered relations with the United States and Japan, along with the identification of the Soviet Union as the primary enemy. It is this transformation that gives depth of meaning to the border crisis, and this is the subject of our study of political change.

Thus, Czechoslovakia provided another turning point in Sino-Soviet relations within the context of the postwar bipolar international system. The Korean War, with its internationalization of the Taiwan issue (and hence of Peking's fundamental interests in national security and completion of the revolution); the watershed period of 1958-1959 in which Moscow's and Peking's divergent perceptions of their respective national interests brought into question the bipolar major premise of their alliance relationship; the internationalization of the Vietnam conflict in 1965, which intensified bipolar pressures to the point of fracturing the Sino-Soviet axis of the socialist camp; and then Czechoslovakia with its transforming effect on adversary relationships—these turning points reflected structural changes that transformed Sino-Soviet relations from an alliance that was a cornerstone of the bipolar system to an adversary relationship of the most fundamental kind, as the border crisis dramatized.

It will be noted that the conceptual context of this study of change owes little to the "deep" historical, ethnological, territorial, and other such sources of the Sino-Soviet conflict that inform many of the studies in this field. Thus, the border crisis is approached in this study as a significant aspect of a political process and the product of political perception and communication within the postwar international system and its conceptual framework. As a review of the recent history of the Sino-Soviet border question will show (Chapter 2), this issue served as one of the political instruments in the burgeoning rivalry between Moscow and Peking. In the early 1960s, both sides used the instrument aggressively, as they did other instruments of political

Introduction

competition, and thus the border issue became inextricabl[y] bedded in a political matrix that not only bore the imprint [of a] long history but was also subject to the contingencies and straints of complex interactions. For this reason, those deep ineluctable forces shaping Chinese and Russian relations over the centuries provide scant analytical insight into evolving political intentions, the ways in which those intentions have been communicated, and the perceptions and misperceptions of the parties to this complex process.

These considerations will become more evident and concrete as this study unfolds. It will be noted that, after the fall from power of N.S. Khrushchev, the border issue became a much less salient instrument of Sino-Soviet rivalry. Khrushchev's aggressive China policy was one aspect of his political style contributing to his ouster, and the post-Khrushchev leadership, as part of its moves to moderate the ideological and political struggle with Peking, sought to submerge the border issue, while at the same time taking care to deploy a strong military presence and capability (a characteristic approach of the Brezhnev leadership) along the China border. This later became an issue in the reinflamed border dispute but, as will be shown, it was a development along Moscow's other front, in East Europe, that became a catalyst for the border crisis that erupted in 1969. Among the foreseeable risks of the invasion of Czechoslovakia, none would seem to have been more remote than what soon developed along the Sino-Soviet border. And still more remote would have seemed the radical change in Sino-American relations that marked the structural shift following Czechoslovakia.

Czechoslovakia and the border crisis conferred a new complexity on Sino-Soviet relations, which now became triangular with the emergence of a Sino-American relationship spanning a gulf that had once seemed unbridgeable. Chinese decision-making, in particular, reflected this added complexity, for the structural shift in the international environment was as much dependent on Chinese internal politics as on actions by the other two powers. In the volatile Chinese political situation, wracked by a series of convulsive struggles generated by the Cultural Revolution, different elements in the Chinese leadership had differing perceptions of events during this period, and this for the very basic reason that the way these events were interpreted and acted upon profoundly affected the destinies of the various leaders and factions. Thus, another layer of complexity and contingency, added to that long present in the interplay of Sino-Soviet international and internal politics, both complicated and, paradoxically, accelerated the evolution of events. Quite apart from genuine conflicts of interest, the incipient Sino-American relationship was beset by perils of misperception, to be expected after two decades of animosity and distrust, and of an unstable Chinese political environment that subjected policy initiatives to what seemed endless challenges. Nonetheless, the impetus from Czechoslovakia and the Sino-Soviet border crisis, combined with the effects of Vietnam war-weariness and of the presence of a right-wing United States administration in reducing American domestic opposition to a new China policy, served to surmount those challenges. Washington undertook to signal a serious interest in exploring a new relationship with China, as Peking had done in a tentative way toward the United States. As will be seen, the Indochina conflict afforded both opportunities and dangers for this process, but the dangers of misperception and overreaction proved to be tractable, given the new conceptual frames in which the emerging opportunities were being viewed. But as will also be seen, a considerable evolution was required in the Chinese political situation before a sufficiently stable mandate could be conferred on the policy line that led to a profound structural change in the international environment.

The Methodological Premises

As a study of political change this is also a study of political communication, of the process of signaling intentions, perceptions, and expectations with varying degrees of distortion, congruence, or dissonance. Failing direct divination of the contents of others' minds, one must infer others' perceptions and intentions from their behavior in its various forms. Epistemologically, the key inferential link in this study is *political signaling*, a form of behavior that bridges not only perception and action but also the familiar (often pernicious) dichotomy of word and deed. Some very salient events, such as the invasion of Czechoslovakia and the shoot-outs on the Sino-Soviet border, figure in this study, but it is the communication of messages in the form of political signaling that is the central subject. Inferences regarding perceptions are based on the messages communicated in the signaling process. That it is a contingent, and therefore inferential, link between perceptions and the messages communicated has negative as well as positive implications, since it permits deception by the source (not to mention unintentional distortion in the communication). But the possibility of deception has a positive side in that it is also a condition for the possibility of political signaling and hence for the possibility of inducing or inviting creative change in the environment (as distinguished from animals' instinctive "signals" that are invariable rather than creative, in an environment in which evolutionary change is accidental rather than intentional). The possibility of deception is the price political man must pay for the ability to initiate change and to reconstitute reality (or to reconstitute the perceptions of those involved in the process, which is a large part of changing the reality).

Signaling is akin to what the philosopher J.L. Austin termed a "performative utterance," one that performs an action by the very fact that the statement is made in a certain context.[3] Thus, an official authorized to conduct a wedding ceremony says to a couple, "I hereby pronounce you man and wife," thereby creating legal ties that entail mutual obligations and expectations. This role must be performed by an authorized official presiding over a duly constituted ceremony, which reflects the contextual meaning of the official's utterance; without the appropriate context, the words would not have the meaning and effect of a performative statement. Similarly, in the case of political signaling, a signal has an internal relation to its context, which is to say that a political signal acquires its meaning *qua* signal from the role it plays in the political process. This is why reference was made earlier to *contextual analysis*, as distinguished from, say, content analysis as well as from hard-headed analytical approaches that insist upon studying "what they do and not what they say." To signal is to act, however arcane or elliptical a form the act of signaling may assume. To signal is to project intentions, influence perceptions, create expectations, and, in myriad ways, to play a role in the political process by shaping new relations among the parties involved.

Thus, it is the purpose of contextual analysis to apprehend the *operative* meaning of political messages, that is, what these messages *do*, what action they perform as signals, the role they play in the political process. This requires a proper appreciation of the context in which such messages acquire their meaning. Contextual factors are especially decisive in determining the meaning of "esoteric communication," in which there is a marked dissociation of first-level textual meaning and second-level signaling meaning. For example, Communists often use historiography in this form of communication, as when the Soviets dust off Leninist scripture purportedly discovered in the archives, or the Chinese elaborate on the history of the Chin dynasty twenty-two centuries ago. It is not ancient times or old manuscripts that are the real subject; life-and-death struggles over power and policy are involved in this signaling process. Virtually any form of discourse can provide the vehicle for this type of communication, which must be interpreted or "decoded" in terms of the political context in which the signaling message has its operative meaning.

To say that a political signal has an internal relation to its context is to say that its operative meaning, logically dependent on the context, cannot be reduced to its textual elements (and thus a reductive form of content analysis—measuring frequency, intensity, and so forth, of the textual elements—would fail to apprehend its operative meaning). Accordingly, a signal is constituted by both the communication as such and the context in which it acquires its operative meaning. Moreover, if signals are constituted by the political context as well as the communication as such, then the other side of this ontological coin is that the political context in turn is constituted by the "objective" reality as such *and* the perceptions and expectations that are the sources of the signals. In other words, this is a political *process*, in which signals create reality by projecting perceptions and expectations that in turn elicit feedback signals affecting the perceptions and expectations that gave rise to the earlier signals.

This study of change contextually analyzes the political communications that signaled the initiatives, new directions, and altered expectations shaping a structural, or systemic, change. It comprises propositions having the logical form, on the basic level, of interpreting the operative meaning of political messages by inferring what they *signal* about the source's intentions, perceptions, and expectations. On a higher logical level, it consists in inferring from the ongoing exchanges of messages—the signaling process— the new trends in policy and perception that constitute political change. Because the signaling process is a critical element in political change and evolution, it should not only be central to the analysis of political change, rather than being regarded as epiphenomenal, but is of special analytical importance in yielding "early warning signals" of the salient, overt, and more directly accessible actions and events that make news and dominate history-writing.

It is an underlying thesis of this study that such signals transmitted in the wake of Czechoslovakia gave semiotic notice of a structural change in the international system. The contextual analysis of this signaling process does not require methodological assumptions about complete rationality of the "political actors" involved, and even less does it assume "unitary" behavior or interests on the level of nation-states in the sense of ignoring the factors of internal factional and bureaucratic politics. To the contrary, it requires appreciation of the interactive character of the process (in both internal and international dimensions), and thus it requires sensitivity to the implications of conflicting and mixed signals, the varying perceptions and priorities of the parties involved, and, above all, the contexts in which the signaling has operative meaning. As noted above, Chinese factional conflict was a critical dimension in the signaling process, and this, in turn, was affected by the signals coming from Washington; moreover, Chinese perceptions of the workings of the Soviet-U.S. relationship have always been a crucial factor in determining China's role in the international system. As for the United States, movement out of the deep-frozen relationship with China that was a legacy of the Korean War was not the product only of a new domestic setting for a new China policy after the 1968 Vietnam Tet trauma and Lyndon Johnson's withdrawal from the presidential race; there were also the perceived opportunities afforded by the eruption of a Sino-Soviet crisis. The United States, accordingly, needed to shape and calibrate its signals with an eye to the variable perceptions of multiple parties to the process.

The Soviet Union was primarily a reactive party, though of course it was no less constrained to deal with the complexity of the process, including the implications for its role in the Communist movement. This study seeks to redress a widespread overemphasis on the military dimension of the Sino-Soviet crisis, one that has been evident not only in a seriously incomplete understanding of the Chinese perception of the meaning of Czechoslovakia but also in a concentration on Soviet shows of force during the crisis. It is safe to assume that the Brezhnev leadership faced pressures from military-minded elements of the Soviet establishment to shake a mailed fist at the Chinese, and, in fact, Moscow took care to send minatory signals reminding them of superior Soviet military might. Nevertheless, the Kremlin was intent on defusing the crisis, and Brezhnev personally identified himself with counsels of restraint, in the interest of fostering an image of a rational, predictable Soviet Union behaving responsibly toward an irrational, belligerent China threatening world order. Thus, in its practice of crisis politics, Moscow was willing to make political concessions to defuse the crisis while seeking to gain political advantages within both the international Communist movement and the international community. Moscow's China policy during this period actually provided the Brezhnev leadership with a broad basis of common purpose, in contrast to the Khrushchev-era situation in which Kremlin factional politics were fed by the hard choices between a détente policy toward the West and a hard-line policy acceptable to the militant wing of the Communist movement. The Brezhnev leadership's uses of the China question in the service of its broader political interests came through in a revealing way at the 1969 Moscow International Communist Conference (Chapter 8).

In sum, contextual analysis of the signaling process does not assume perfect rationality (or imply any particular degree of irrationality) on the part of those involved, nor does it presuppose unitary interests at any particular level of analysis (even at the level of individuals, signaling behavior must be interpreted contextually in the light of their changing opportunities and vulnerabilities). To determine the degree of decision-making conflict in a factional context requires an analysis of just that, the context; in the Sino-Soviet and Sino-U.S. relationships developing during the period of this study, there was a highly charged factional context in Chinese decision-making, much less so in the Soviet and United States political situations. But since the process is open-ended and inherently political, and hence reversible, it behooves us to attune our analytical antennas to early warning signals of incipient change in new directions. Changes tend to be incremental (even if at times spectacular and newsworthy) after a critical shift, but the early phases of structural change are highly contingent, characteristically delicate, and vulnerable to misperception, or, more typically, missed perception. The ultimate purpose of a study such as this is to heighten appreciation of the dynamics of the signaling process in order to be in a position, conceptually and analytically, to apprehend the early signals of structural change.

TWO

Politics of the Border Question

The Sino-Soviet border conflict is deeply embedded in a geopolitical, historical, and ethnological matrix. These conditioning factors have lent themselves to analyses plumbing the depths of the conflict in such a way as to show a grand sweep of historical inevitability. Whatever intellectual uplift might be derived from such bold interpretative designs, a close contextual analysis, within bilateral and international as well as domestic contexts, is required to elucidate the political decision-making and signaling that have determined the timing and character of the events in which the conditioning factors have been given expression. Politics, as they say, is the art of the possible, and the possibilities are created and limited by historical, geographical, and other such considerations. Our concern here is with the political uses of those possibilities, which is to say, a concern with the border issue as an instrument of the political process. A review of the uses of this instrument not only aids in an understanding of how the Sino-Soviet border crisis arose but also elucidates the meaning of the Sino-Soviet conflict in itself.

As the Soviets have frequently pointed out in their effort to document the expansionist tendencies of Maoist China, Mao was already saying in the 1930s that Outer Mongolia would "automatically" become part of China after the Communist revolution succeeded in China.[1] The question of Mongolia also figured in the

1950s at times when the Soviet leadership situation was unsettled and the Chinese deemed the time ripe to introduce an irredentist issue. Thus, when Khrushchev and Bulganin went to China in 1954, the year after Stalin's death and thus at a time when Khrushchev was seeking leverage in his move toward ascendancy in the Kremlin, the Chinese "took up" the Mongolian question with the guests; but the guests, according to Mao, "refused to talk with us."[2] The visitors did in fact make conciliatory gestures in courting Peking, as in agreeing to Soviet withdrawal from the Port Arthur naval base and the revocation of the Stalinist 1950 arrangement for joint exploration of Sinkiang mineral resources; however, it was not surprising, in view of Moscow's acute sensitivity to irredentist claims, that Mao's attempt to discuss Mongolia's status was rebuffed. According to the Soviet version, the Chinese had offered to reach agreement with Khrushchev on making Outer Mongolia into a Chinese province.[3] On still another occasion, in January 1957 (a few months before the climax of a Kremlin power struggle), Chou En-lai apparently suggested to Khrushchev that steps be taken toward arranging a settlement of territorial issues concerning China, Japan, and other areas. According to Chou's subsequent account, he did not receive a satisfactory reply.

Notwithstanding these signs of a Chinese interest in seeing their country's boundaries moved out to what they might regard as historical completeness (the interest dictating their moves to secure control over Tibet), the Mongolian question has not figured centrally in the Sino-Soviet territorial question. In contrast to Peking's adamant refusal to countenance or sanction any *de jure* compromise of its claim to sovereignty over Tibet or Taiwan, diplomatic relations were established between the Chinese Communist regime and the Mongolian People's Republic, a nominally independent Soviet satellite. Peking's approach to the MPR has been similar to its approach to other Communist states in the course of the Sino-Soviet rivalry, a varying mixture of inducements to lessen dependence on Moscow and of political and ideological attacks on pro-Soviet relationships. In this sense the MPR has figured in the Sino-Soviet border dispute in terms of the outer reach of Peking's position, that is, in connection with the threat to reopen the whole territorial question and thus to stimulate a chain reaction of irredentist grievances on the part of a range of countries neighboring the Soviet Union in Europe and Asia. To be sure, Peking's residual claim to Outer Mongolia would be a part of that irredentist threat, but, to repeat, the threat itself was not central to the border dispute and represented more an inducement to negotiations than an issue of negotiations.

According to the accounts by both sides, border troubles as a direct and persistent influence in Sino-Soviet relations developed in the 1960s as a concomitant of the evolving rivalry produced by the failure of mutual expectations in the previous decade of alliance and cooperation. In the Chinese version, Soviet provocations along the border were one of the means, like the recall of Soviet technical experts, to which Moscow resorted to pressure the Chinese into line after the traumatic early collision of Soviet and Chinese leaders in Bucharest in June 1960. After the 22nd CPSU Congress in late 1961, at which Khrushchev presaged stronger pressures against Peking by his vigorous attack on its Albanian surrogate, the Soviets undertook what the Chinese said were further steps damaging to Sino-Soviet relations; among these was the rankling incident in Sinkiang in the spring of 1962 involving the flight of tens of thousands of members of the national minorities into the Soviet Union. The Soviet version of the emergence of border problems conforms in essential respects with the Chinese account, particularly in dating the origins to 1960 and thus, in effect, confirming that the border dispute was both an effect and an instrument of the political rivalry burgeoning during that period. Significantly, in addition to the 1962 Sinkiang incident, the Soviets have pointed to moves taken by the Chinese already in the early 1960s to contest Soviet jurisdiction over islands in the Amur and Ussuri. If the known pattern of Khrushchev's aggressive behavior in trying to pressure the Chinese into line lends credence to Peking's account of the early border troubles, so also would Chinese behavior later in the decade seem to confirm the Soviet charge of deliberate Chinese moves to challenge Soviet possession of islands in the border rivers. In short, both sides were using the border as a weapon in their political struggle.[4]

It was after the Cuban missile crisis of autumn 1962 that the border question began to emerge more explicitly as a factor in the signaling process through which the political struggle was being expressed. In a major address to the Supreme Soviet in December 1962 defending Soviet behavior in that crisis against dual Chinese charges of adventurism for installing the missiles and capitulationism for bowing to U.S. demands for their removal, Khrushchev tauntingly mentioned Chinese acquiescence in the British colony of Hong Kong and the Portuguese colony of Macao in contrast to Indian action in seizing Goa from the Portuguese and the Indonesians' expulsion of the Dutch from New Guinea.[5] A Soviet client party, the U.S. Communist Party, made the point more directly in a statement the next month, asking why the Chinese took a "double-standard approach" to Taiwan, Hong Kong, and Macao by refusing (and "correctly" so, the CPUSA said) to take the adventurist path "that they advocate for others."[6]

At this stage, Moscow and Peking were still observing the protocol of conducting their polemical exchanges indirectly, and so the Chinese addressed their reply to the taunt by the CPUSA rather than to its patron. A *People's Daily* editorial on 8 March 1963, "A Comment on the Statement of the Communist Party of the USA," let it be known that the Chinese knew "from what quarter" the CPUSA "learned this ridiculous charge" and that they knew "the purpose of the person [Khrushchev] who manufactured it." Inasmuch as "some persons" had mentioned Taiwan, Hong Kong, and Macao, the editorial explained, the Chinese were "obliged to discuss a little of the history of imperialist aggression against China," and in this context the editorial mentioned three of the nineteenth-century treaties on the Sino-Russian border in reciting the series of "unequal treaties" imposed on China in the past century. After stating Peking's policy calling for peaceful negotiation of border questions and for maintenance of the status quo pending a settlement, the editorial introduced an ominous if implicit threat to the Kremlin in noting that "certain persons" had "picked up a stone from a cesspool" after the Cuban crisis: "But whom has this filthy stone actually hit? You are not unaware that such questions as those of Hong Kong and Macao relate to the category of unequal treaties left over by history, treaties which the imperialists imposed on China. It may be asked: In raising questions of this kind, do you intend to raise all the questions of unequal treaties and have a general settlement? Has it ever entered your heads what the consequences will be? Can you seriously believe that this will do you any good?"[7]

Having thus emerged in a polemical context of thrust and riposte, the border issue became an element in the ideological campaigns that exploded in full view and on a large scale in mid-1963. The first in the monumental series of Chinese commentaries on the 14 July CPSU Open Letter[8] traced the origins of border troubles as part of the genesis and development of the Sino-Soviet conflict. In addition to explaining that border provocations were part of Moscow's arsenal of pressure weapons after the 1960 Bucharest conference, the Chinese account blamed Soviet organs in Sinkiang for having fomented the mass exodus of refugees from the Ili area and charged that the Soviets had refused Peking's repeated demands for the repatriation of the refugees. Reflecting sensitivity to its vulnerabilities in this region, Peking called this "an astounding event, unheard of in the relations between two socialist countries."[9]

Two weeks after the Chinese published their account of the origins of the Sino-Soviet conflict, Moscow issued a statement which was a part of the ongoing exchange over the nuclear test ban treaty signed that summer but which contained a passage on the border situation responsive to the Chinese commentary. In a charge mirroring the Chinese account, the Soviet statement claimed that the Chinese had been "systematically violating" the border since 1960 and had been guilty of more than 5,000 violations in 1962 alone. To document the charge that the Chinese had been attempting to contest Soviet jurisdiction along the riverine borders, the statement quoted from instructions said to have been found on a Chinese intruder, calling for the Chinese to fish on the disputed islands and to tell Soviet border guards that these islands belonged to China. The statement disclosed that Moscow had on a number of occasions suggested that consultations be held on demarcation of separate stretches of the border to exclude any possibility of misunderstanding but that the Chinese had evaded this proposal. It took note of the Chinese threat in March to open the border question, remarking that "Chinese propaganda is giving definite hints that there has been unjust demarcation of some sections of the Sino-Soviet border in the past." Signaling Moscow's determination to head off any further development of this issue, the statement sternly warned that the "artificial creation" of territorial problems, "especially between socialist countries," would amount to entering on a "very dangerous path."[10] Chinese taunting on this issue had clearly touched a very sensitive nerve.

Also at this time, the Soviets chose to present their version of the 1962 incident in Sinkiang that so rankled the Chinese and had dramatized the potential for troublemaking among the restive minority peoples of Central Asia straddling the state border of the rival countries. Using a local press organ, the *Kazakhstanskaya Pravda*, as the channel, the Soviets explained that the exodus into the USSR from Sinkiang in the previous year was sparked by a "monstrous" incident in which a number of residents of the Ili Kazakh district went to the local Communist Party office seeking permission to emigrate and were laced by machine-gun fire. One of the authors of these accounts, Zunun Taipov, a former major general in the Chinese army who had fled to the Soviet Union, expressed bitterness over the "incredible persecution and repression" of his "brothers," the Uighurs and Kazakhs, Kirgiz and Mongols, living in Sinkiang under Chinese domination. The use of Taipov as a spokesman at this time and again during the severely troubled late-1960s was particularly ominous in that he might serve as a standard-bearer for a call to armed resistance in the name of a "national liberation movement" by the non-Chinese peoples in Sinkiang.[11] The inflammatory propaganda carried in the Soviet Central Asian press was amplified in broadcasts from transmitters in that region that could be heard in Sinkiang in the languages of the national minorities there. The effect of this propaganda effort was indirectly confirmed in the warnings by Chinese authorities about "large-scale subversive activities" by the Soviets directed against Sinkiang. Thus, a session of the Sinkiang People's Congress in April 1964 reiterated the charge that the Soviets had "enticed and coerced tens of thousands of Chinese citizens into going to the Soviet Union" and had used radio transmitters to attack the CCP leadership and distort the history of Sinkiang "in an attempt to undermine the unity of the Chinese people of various nationalities." The use of this kind of provocative propaganda was thus both a symptom and a cause of trouble in the Sino-Soviet relationship, as indicated by its activation during periods of deteriorating relations.[12]

The Soviet statement of 21 September 1963[13] referred to Moscow's overtures in private communications regarding "consultations" on demarcation of the border, and now another effort was made as part of the two sides' exchanges on ways of improving their strained relations. A CPSU letter of 29 November 1963, not released publicly until five months later,[14] noted that the Chinese, too, had recently spoken in favor of resolving the border dispute. The letter indicated, however, that Moscow was not interested in entering into a competitive reappraisal of the historical background: the Soviets would not defend the tsars, it said, nor surely would the Chinese defend the Chinese emperors. But "we cannot disregard the fact that historically formed boundaries" now exist, the letter stressed. In addition to this private communication to the CCP, Moscow turned to the international community in its effort to cut short any move to exploit border issues as an instrument of political conflict. On the last day of 1963, Khrushchev addressed a message to heads of state or government, proposing the conclusion of an international agreement on the renunciation of force for the settlement of territorial disputes or border questions.[15] As a status quo power having a strong interest in ratification of the territorial settlements emerging out of World War II in particular, Moscow has characteristically promoted renunciation-of-force agreements as a means of consolidating its territorial gains. In its application to the Sino-Soviet border situation, Khrushchev's proposal was intended to remove from Peking's hand the threat to reopen the inherited territorial arrangements by any means of its choosing, including the use of force (as ultimately took place at the end of the decade).

The Sino-Soviet border discussions began on 25 February 1964 and provided a forum for testing the prospects of reducing the strains that had flared up dramatically in the past year. Four days later a CCP letter replying to the CPSU's communication of the previous November set forth Peking's position on the border question.[16] It reaffirmed the stand that, although the old treaties relating to the Sino-Soviet border were unequal treaties, the PRC was "nevertheless willing to respect them and take them as the basis for a reasonable settlement." It expressed the view that the issue could be settled through negotiations and that, "pending such a settlement, the status quo on the border should be maintained"; but it went on to charge that, in recent years, the Soviets had "made frequent breaches of the status quo on the border," occupied Chinese territory, and provoked border incidents. The charge of Soviet occupation of Chinese territory seriously complicated reaching agreement on maintenance of the status quo in that it begged the question of what had to be negotiated in a settlement. "Still more serious" in Peking's view was the Soviet practice of "large-scale subversive activities in Chinese frontier areas, trying to sow discord among China's nationalities by means of the press and wireless, inciting China's minority nationalities to break away from their motherlands, and inveigling and coercing tens of thousands of Chinese citizens into the Soviet Union." This was a very revealing acknowledgment of the potential for Soviet troublemaking in China's far-western marches, and the recurrence at that time of Peking's complaint on this score made this aspect of the border dispute a more central issue than at the time of the clashes at the end of the decade. This was probably due in large measure to the serious disaffection toward the regime caused by the severe economic dislocations that followed the

Great Leap Forward campaign. The exodus from Sinkiang coincided with a rush of refugees into Hong Kong that reflected the privations being suffered in China. Acute economic ills combined with other grievances of national minorities would have made Sinkiang a vulnerable target for Soviet fishing in troubled waters, particularly if grievances had been exacerbated by a cutoff of emigration by Peking after its experimenting with an open door policy as a means of relieving the pressures of popular discontent. Such a sequence of developments would help explain the explosive events in the Ili district in 1962 and would lend credence to Moscow's account of petitioners seeking to emigrate.

Peking used a major policy statement by Chou En-lai in April 1964 to respond to Khrushchev's appeal to the international community for a renunciation of force in border disputes. In a report to the National People's Congress Standing Committee on his recent foreign excursion,[17] Chou derided Khrushchev's proposal as "a new fraud" which "deliberately confused imperialist aggression and occupation of other countries' territories with territorial questions between nations left over by history." Chou said that boundary questions between African and Asian countries (a primary audience for Chou's statement) should be settled through peaceful consultations, and he added that this also holds between socialist countries (the Sino-Soviet border dispute being one of those said by Peking to be left over by history); but he stressed that "imperialist aggression and occupation of other countries' territories was a matter of a completely different nature." The aggrieved countries "naturally had every right to recover their lost territories by any means, and thus they could not be asked to renounce the use of force in all circumstances."[18] Significantly, Chou's discussion of Khrushchev's proposal followed a passage on United States "occupation of China's territory of Taiwan." The distinction Chou made between different categories of territorial disputes underlay one of Peking's most fundamental diplomatic positions, namely, its resolute rejection of any undertaking that might compromise its claimed right to use force if necessary to effect the "liberation" of Taiwan from the Chinese

Nationalists, a matter Peking regards as its internal affair. This position stood at the center of the Sino-American deadlock, and now it was becoming involved in the Sino-Soviet border dispute.

Chou's rebuttal of the Khrushchev proposal was logically deficient, raising the possibility that Peking was concerned to avoid a renunciation-of-force commitment because of its implications for the Sino-Soviet dispute as much as for the Taiwan question. In his proposal, Khrushchev had, in fact, distinguished territorial disputes arising from "already formed and established state borders," as one category, from situations in which states have not yet been able to free from colonial control all the territory rightfully belonging to them; and indeed he had named Taiwan first as an example of the latter category. He thus had explicitly excluded the Taiwan question from the application of his proposal on renunciation of force, and Chou's reply was, strictly speaking, inapposite. Moscow might well have concluded from this exchange that the Chinese were mainly concerned at the time to leave themselves free to use force to contest Soviet jurisdiction over disputed areas, a conclusion the more likely in that Khrushchev's proposal had arisen in the context of the Sino-Soviet border conflict.

If Chou's position might have aroused Soviet suspicions, no less a figure than Mao himself was soon to detonate a blockbuster that would undermine the border talks and exacerbate Sino-Soviet abrasions to their sharpest point up to that time. In an interview published in the Japanese press but not carried in the Chinese media, Mao told a visiting Japanese Socialist Party delegation on 10 July 1964 that there were "too many places occupied by the Soviet Union," and he proceeded to enumerate a wide-ranging series of territorial settlements establishing the Soviet Union's current borders. Mao's sweeping indictment of these territorial settlements was cued by a Japanese question about the loss by Japan of the Kuriles as part of the post-World War II settlement, and most of Mao's examples related to the postwar dispositions in Eastern Europe. But he went on to note that a century ago the area to the east of Lake Baykal became Russian territory, and here he delivered the main thrust of his provocative message: "We have not yet presented our account for this list" of lost territories.[19] Mao's move was undoubtedly a reaction to the foundering border talks and represented the ploy of threatening a reopening of the whole territorial question in the face of Soviet intransigence. It was this ploy which the Chinese were to recall in March 1969 when they pointed out that, in the 1964 talks, they had warned they would have to "reconsider" their position "as a whole" if the Soviets refused to acknowledge that the inherited treaties were unequal and to take those treaties as the basis for determining the boundary. This threat was made the more menacing by Mao's linkage of Chinese grievances with a host of other nations' real or potential senses of grievance, the very sort of chain reaction the Kremlin most feared. He was quite explicit on the issue of the Kuriles, which the Japanese call their "northern territories" and which Mao said "must be returned to Japan."

Mao's ominous remarks had a shattering effect on Moscow, and the border talks were suspended the following month, not to be resumed until the end of the decade (and then on terms closer to Peking's conception). Moscow was goaded into a political offensive on the border question to counteract the effects of Mao's sweeping charges. An authoritative *Pravda* editorial article on 2 September 1964, "In Connection with Mao Tse-tung's Talk with a Group of Japanese Socialists,"[20] acknowledged the scope of Mao's charges by noting that he was not only making claims on Soviet territory but was "portraying his claims as a part of some 'general territorial question.' We are faced with an openly expansionist program with far-reaching pretensions." *Pravda* was concerned to discredit the method of using historical references to advance territorial claims, pointing out that anything can be proven by using this method, depending on which historical period is cited. Moscow's argument was that, since Soviet people had lived and worked on the land whose inclusion in their country was being questioned, the existing border had "developed historically and was fixed by life itself." Noting that the early

Soviet regime had annulled all unequal treaties with China and that later the Soviets had given up the Port Arthur naval base, *Pravda* recalled that Lenin had condemned the tsarist government's seizure of Port Arthur but that "it was none other than Lenin who said: 'Vladivostok is far away, but this town is ours.'" As for the chain-reaction effect of irredentist claims, *Pravda* observed that Mao had "crossed out with amazing ease the entire system of international agreements" concluded after World War II, and it professed to find it "difficult to believe that he is unaware of the most dangerous consequences that could result from any attempt to recarve the map of the world under present conditions."

In addition to the major *Pravda* editorial article, Khrushchev typically personalized the debate by arranging an interview of his own with a Japanese delegation. He argued that there must be a single standard for judging all aggressors of the past, whether Russian tsars or Chinese emperors, both of whom waged wars of aggrandizement. Rather provocatively (and again typically), Khrushchev cited the sensitive case of Sinkiang, noting that the indigenous population differs sharply from the Chinese ethnically, linguistically, and otherwise, and that they had been conquered by the Chinese emperors. He noted further that both the Soviet Union and China are multinational states that took shape as a result of historical processes; the bulk of the Kazakh people live in the Soviet Union but some of them live in China; likewise in the case of the Kirghiz people. The implication was that both sides should leave well enough alone, but the Soviets could also play the irredentist game, and Sinkiang was the most vulnerable target.[21]

As still another prong of Moscow's political offensive, Foreign Minister Gromyko proposed to the U.N. Secretary General that a Soviet resolution, "On the Renunciation by States of the Use of Force for Settling Territorial and Frontier Disputes," be included on the agenda for that autumn's session of the General Assembly.[22] That this was a renewal, in a context of new urgency, of Khrushchev's proposal of the previous December was indicated by Gromyko's reference to the earlier message as having "exhaustively clarified" the considerations underlying the Soviet initiative. Pursuing still another diplomatic channel, Moscow proposed to the Chinese on 28 September that the border discussions be resumed on 15 October. But Khrushchev was deposed in the meantime, and a new cycle in the Sino-Soviet relationship was begun. The border dispute and Khrushchev's plans for an international party conference, two issues on which the Sino-Soviet conflict had been focused, were for the time being put in abeyance. By the end of the decade, however, those issues had again come to a head.

In the course of the Sino-Soviet border dispute the Chinese have made a point of saying that they have managed to arrange satisfactory boundary agreements with all their neighbors except with the Soviets and "the reactionary nationalists of India."[23] That observation naturally prompts the question of the degree to which Peking's disputes with those two big neighbors were similar in origin and character. Merely raising the question in those terms suggests one answer, that the very fact that the two big neighbors became embroiled in border disputes with China derived from Peking's perception of them as natural rivals owing to their sheer size. Such an answer does not, however, seem to account for the Sino-Indian conflict, and hence the answer is defective as a general covering explanation for Peking's border disputes. Though the Sino-Indian and Sino-Soviet relationships enjoyed their halcyon times during the same period, in the mid-1950s, the border dispute between India and China was much more specific in its causality and less a symptom of other conflicts of interests than was the Sino-Soviet rift. Peking's concern to secure its control over Tibet was, from the beginning, an irritant and a potential source of major trouble between the newly independent India and the Communist regime in China. The Chinese road connecting Sinkiang and Tibet was designed to serve the objective of exercising control over distant borderlands, and the discovery of that road in the Aksai Chin area along the western Sino-Indian border, combined with Peking's heightened concern over its control over Tibet after the major revolt there in 1959, lay at the heart of the

Sino-Indian dispute. The surgical precision of Peking's diplomatic and military moves in that dispute attests to the specificity of Peking's interests in the Sino-Indian conflict.

Whereas the Sino-Indian border dispute was essentially a by-product of Peking's concern to secure control over Tibet, the Sino-Soviet border troubles were part and parcel of the fundamental rivalry that emerged from the conflict of interests that became manifest in the late 1950s. As Peking tells it, border troubles were the result of Moscow's attempt—indeed were one of the instruments of that attempt—to pressure the Chinese into line after the latter had mounted a challenge to Moscow's authority in 1960. In the Soviet version, the border problem was one of the manifestations of Peking's effort to impose its strategy and priorities on the Communist movement, an effort signaled by the Lenin anniversary polemics of April 1960 and reflected across a wide range of challenges to the Soviets including the border question. As shown by the uses of the border question (what might be called the politics of the border question) in its different phases of development, notably in 1963-1964 and again after the Soviet invasion of Czechoslovakia in 1968, it was not essentially a territorial question as such but one among other arenas in which the political struggle between the two sides could be conducted. In other words, the Sino-Soviet border issue has been one of the forms, as has been (for example) the question of peaceful coexistence, in which the Chinese challenge and the Soviet response have taken shape. The root of the overall conflict had been the two sides' divergent perceptions of the international system and of its implications for their divergent priorities. As is argued in this study, Peking's perception of the implications of the invasion of Czechoslovakia set the context for the severe border crisis that emerged the following year.

In one key respect, the positions taken by India and the Soviet Union in their respective border disputes with China have been strikingly similar. Where Peking in each case insisted on a need for "overall negotiations" to produce new border treaties and thus to erase grievances "left over from history," the other side, in each case, was concerned to avoid opening the door to renegotiation of the borders, lest the Chinese should raise sweeping irredentist claims of a dangerously destabilizing nature. Thus, both India and the Soviet Union offered to hold consultations with China on a more precise demarcation of various stretches of the border, but only in such a way as to avoid opening the border to renegotiation. The logic of their argumentation led the Indians and the Soviets into similar semantic distinctions. For Nehru, the difference was between negotiations and talks; for Moscow, between negotiations and consultations; in each case the distinction served the purpose of resisting Peking's demand for negotiating new border treaties.

Peking demanded a renegotiation of the inherited Sino-Soviet border treaties because they were "unequal" in their origins. In the Sino-Indian dispute, Peking's position was that no predecessor Chinese government had ever signed the McMahon Line delimiting the boundary in the east. But, despite that formal difference, Peking followed a similar substantive approach in each case by expressing its readiness to negotiate new treaties on the basis of the "unequal" Sino-Soviet treaties in the one case, and on the basis of the unilateral McMahon Line in the other.[24] There has been a substantive difference in Peking's treatment of the question of the status quo in the two disputes. Though, in each case, the Chinese offered to maintain the status quo pending a comprehensive settlement, they have been inclined to introduce a juridical aspect to their definition of the Sino-Soviet border, much as the Indians did in the Sino-Indian dispute. In that dispute, Peking made much of observance of the line of actual control as the de facto boundary pending a negotiated settlement. But India, while also professing agreement to maintenance of the status quo, demanded in effect the restoration of the status quo ante, not the line of actual control;[25] the Indian demand would have required the Chinese to evacuate the Aksai Chin area in the western section of the border, thus abandoning the road that lay at the heart of Peking's concern in the whole dispute. In the Sino-Soviet situation, there has been an element of ambiguity in Peking's position as regards the status quo. While calling for maintenance of the status quo and opposing any moves to advance the line of actual control, Peking added the stipulation that, in sections where a river forms the boundary, neither side should cross the central channel. This stipulation amounts to a qualification or an addition to the principle of maintenance of the status quo by observance of the line of actual control as the de facto boundary; for, if the Soviets were in actual control of an island lying on the Chinese side of the central navigation channel, they would be required under this provision to abandon that control. It can thus be understood that Peking's fateful decision to contest Soviet control over border river islands served the purpose of bringing the de jure and de facto aspects of its negotiating position into harmony. But such a decision was hardly conducive to maintaining the status quo along the border, as the border crisis of 1969 dramatically confirmed.

THREE

Impact of Czechoslovakia, 1968

Seen in historical and global perspective, the effects of the Soviet-led invasion of Czechoslovakia in August 1968 present an intriguing paradox. At the time of the invasion it was perceived, in Moscow as well as elsewhere, as entailing grave risks on a broader international scale, risks that the Kremlin had to weigh against those it perceived in the seemingly uncontrollable forces of ferment being unleashed by the Czechoslovak reform movement. The international risks (excluding the unlikely event of Western military counteraction) were essentially two: a possible undermining of Moscow's cherished project of an International Communist Conference, scheduled to convene in Moscow in November that year; and a dangerous poisoning of the atmosphere surrounding East-West relations, particularly from the diplomatic fallout from the military intervention in Czechoslovakia. Risks of the latter sort were most sharply focused by the diplomatic moves toward opening negotiations on the limitation of strategic arms, an announcement of which had been scheduled for the very day after the launching of the invasion.

In the event, and here lies one aspect of the paradox, the Soviets were remarkably successful in managing these risks. To be sure, the International Party Conference was postponed until the following spring, and strategic arms talks with the United States were not put back on track until over a year later. But these were mere delays, and minor ones considering the long history of East-West arms control diplomacy and in view of the fact that the Communist Conference project had been initiated as far back as November 1966.[1] Meanwhile, also, the Soviets succeeded in installing an amenable regime in Prague that avoided a provocative return to the Stalinist Novotny era while eventually acceding to Moscow's demands for conformity and acknowledgment of the legitimacy of the intervention. All in all, it appeared to be a model (from Moscow's standpoint) of risk management. Things were "normalized" in what had been an acceleratingly restive Czechoslovakia, and things were soon back to normal in the international community.

All in all, however, things were back to normal only on the surface of global affairs. Even at the surface level, the extraordinary lengths the Sino-Soviet conflict reached with the border fighting in 1969 could be interpreted within the framework of developments dating from the beginning of the decade and the increasingly overt nature of that conflict. But the exceptionally significant development, and here is to be found the other aspect of the paradox, was that the impact of the invasion of Czechoslovakia, an impact concentrated by developments in the Vietnam conflict, produced a geologic shift in the global political system. To put these two aspects together, the paradox lies in the fact that the very "normality" of the intervention in Czechoslovakia induced a pattern of reaction that was to restructure the international system that had prevailed for more than two decades of postwar global politics. Viewed in these terms, the Sino-Soviet border fighting in 1969 should be interpreted as symptomatic of this geologic change, an eruption on the surface registering the underlying shifts.

The key to the paradox lies in Peking; more particularly, in Peking's acute sensitivity to implications for China of developments elsewhere in the international system. This sensitivity had long been evident, particularly in its corrosive effect on the Sino-Soviet relationship. The 1958 Taiwan Straits crisis, the 1959 Sino-Indian border conflict, the simmering tensions between Peking and

Moscow that flared into ideological rivalry in the 1960s—these developments were essentially affected by Peking's concern over the effects on its vital interests of the evolving moves by Moscow and Washington to stabilize their relationship in a spirit of détente. Peking's enduring interests were rooted in its concern over the security of its borders and its desire to complete the Chinese Communist Revolution by eliminating the rival regime entrenched on Taiwan. But these interests were by necessity—the Korean War decided that, if nothing else—subject to the interplay and tides of global politics, and hence these vital interests were embedded in the workings of the international system.

By the mid-1960s, after a climactic debate over how to react to the expanding Vietnam conflict,[2] the Chinese leadership defiantly permitted the Sino-Soviet alliance to become defunct. That decision was the more momentous in that it undercut one of the deterrent shields protecting China and exposed that militarily inferior country to the contingencies posed by increasingly assertive American power in Vietnam. Peking's perception of the international system, as it was operating most saliently at that time in Vietnam, was summed up in the formulation that Washington and Moscow were seeking to draw the Indochina question into "the orbit of U.S.-Soviet collaboration." There was an obvious continuity between that perception and that which underlay Chinese concern in the late 1950s over moves toward U.S.-Soviet détente. Its efforts to scuttle the moves toward détente having proven unavailing, Peking now opted to withdraw as much as possible from the international system behind a smoke screen of Maoist revolutionary propaganda. The decision to allow the Sino-Soviet alliance to lapse was part and parcel of this withdrawal from the international system.

Though events moved swiftly and Sino-Soviet relations in particular were to undergo dramatic deterioration (as manifested, for example, by the marathon siege of the Soviet embassy in Peking in early 1967), the logic of these events remained consistent over time; there was an underlying continuity in Peking's perception of the Vietnam conflict as a test of the dominant international system, one in which the congruence of Soviet and U.S. interests overrode the strains produced by that conflict. Accordingly, during that period, Peking's persistent, indeed fanatical, purpose was to act as a spoiler in attempting to prevent the international system from working. Above all else, that meant preventing a settlement in Vietnam, for such an outcome would mean a vindication of the international system and the removal of an irritant in U.S.-Soviet relations. As infantile as Chinese behavior—antics would be a better term—seemed at that time, a consideration of the implications of a Vietnam settlement for Peking's own interests reveals a logic that even the wild excesses of the Cultural Revolution should not be allowed to obscure. A Vietnam settlement would have meant that a drive to complete a nationalist Communist revolution in an Asian country had been thwarted by American intervention; and that, moreover, the Soviet Union was willing to acquiesce in its ally's frustration in the larger interests of maintaining the stability of the international system. This consideration relates to Peking's acute concern over the sacrifice of its own vital interests in the Taiwan question to the maintenance of Soviet-U.S. equilibrium.

Such was the situation prevailing at the time of the August 1968 invasion of Czechoslovakia. Already that year, a significant change in the Vietnam conflict had taken place with President Johnson's announcement in March of a ceiling on the number of American troops there, followed by an agreement to open negotiations. Peking, however, had reacted to that change in a manner fully consistent with its previous policy, denigrating Johnson's announcement as a "plot" and insisting that the U.S. desire for negotiations was "a big fraud."[3] Thus, developments in Vietnam as such were not the cause of the geologic shift to come. One must look to the impact of Czechoslovakia on Peking's perceptions to find the sources of that shift.

The paradox mentioned above—the fact that the invasion of Czechoslovakia served to confirm the international system but, in so doing, resulted in a transformation of the system—calls for a distinction between the impact and implications of Czechoslovakia, on one level, and, on a second level, the interplay of reactions to that impact and its implications. Put another way, this is a distinction between reactions to the invasion and reactions to those reactions. The shock waves reverberating through the Communist movement from the invasion brought into sharper relief the tensions straining Communist ranks. In considerable measure, these reverberations amplified trends set in motion in the middle of the decade by the effects of the escalating hostilities in Vietnam and the post-Khrushchev Soviet line of united action in support of the Vietnamese comrades (a "sham" support, the Chinese insisted in the course of repudiating Moscow's line). Those trends had significantly reversed—at Peking's expense—the polarizing effects in the Communist movement produced earlier in the decade by Peking's ideological challenge to Moscow's authority. But now the amplification of these trends reached a point where quantitative change turned into qualitative change, and here Peking's perceptions in the wake of Czechoslovakia were crucial. The invasion produced such strange bedfellows as the Yugoslavs and their bitter enemies in Albania, while underscoring the divergences between the Chinese and the other militant regimes in North Vietnam, North Korea, and Cuba. The depth of Czechoslovakia's impact can perhaps best be measured by an unprecedented rebuke administered to that sacred cow of world communism at that time, the Democratic Republic of (North) Vietnam, which piqued the Albanians by approving what Tirana called the "barbarous fascist-type aggression" committed against Czechoslovakia.

Apart from the participating Warsaw Pact allies, endorsements of the invasion were forthcoming from the Soviet satellite in Outer Mongolia, from traditionally docile pro-Soviet parties in Latin America, and from the usual scattering of client parties revolving in the Soviet orbit. Probably the most rewarding support for Moscow was that from Hanoi, Pyongyang, and, in a qualified manner, Havana, radical independents with which the Soviets had had serious ideological and political differences. Those embattled regimes viewed intervention in Czechoslovakia as a measure necessary to preserve the integrity of "the socialist camp," an alliance system in which they had strong vested interests. Ever the maverick, Fidel Castro managed both to endorse the invasion and to chide the Soviets, using the occasion to indulge his own grievances against Moscow while approving the intervention in notably candid terms of realpolitik.

At the other end of the spectrum, the independent-minded Yugoslavs and Romanians saw their interests jeopardized by the use of Soviet force to crush an exercise in limited autonomy within the Soviet sphere and to impose Moscow's norms of Communist behavior on East Europe. The recriminations arising from criticism of the invasion leveled by Belgrade and Bucharest provoked a renewal of harsh Soviet polemics with the Yugoslavs and elicited direct attacks by Moscow on Romanian positions (regarding the Middle East and nuclear proliferation, for example) which had previously been discreetly ignored in public or criticized only indirectly. If the independent radicals welcomed the invasion as serving to confirm the global division between the socialist and imperialist camps, the Yugoslavs and Romanians, outflanking Moscow in seeking to promote détente, feared just what the Vietnamese and Korean Communists welcomed.

The Romanians held an emergency meeting on 21 August, the first day of the Soviet occupation of Czechoslovakia, and party leader Nicolae Ceausescu delivered a blistering denunciation of the invasion as a "great error and a serious danger to peace in Europe and to the fate of socialism in the world . . . There is no justification whatsoever and no excuse can be accepted for considering even for a moment the idea of military intervention in the affairs of a fraternal socialist state." The Romanians reinforced this stiff message to Moscow by signaling that they would offer armed resistance to any such move directed at them: Ceausescu declared that the emergency meeting had decided to establish worker-peasant defense units to protect "the independence and sovereignty of our socialist fatherland." This was likely to receive a sympathetic hearing from Romania's Yugoslav neighbors, with their history of guerrilla resistance to the Nazi invaders and their defiance of Stalinist threats. And indeed, Romanian and Yugoslav readiness to stand up to the Soviets with arms in hand received a good notice from the fiercely militant Albanians, thereby forming a Balkan phalanx which, reinforced by Chinese declarations of support, must have complicated any calculations anyone in Moscow may have been making about the cost of conducting a clean sweep of the mess in East Europe now that the military option had already been exercised.[4]

Along with Yugoslavia and Romania at that end of the Communist spectrum favoring a relaxation of East-West tensions, the West Europe parties were almost solidly in opposition to the invasion. The two most important, the Italian and French CPs, went on record on 21 August with Politburo statements deploring the intervention. Characteristically, however, the French Communists still found it difficult to sustain a show of independence of Moscow and early on gave signs of wanting to return to the Kremlin's maternal bosom. Already on 22 August, a French CP Central Committee statement endorsing the previous day's Politburo statement made a point of pledging cooperation with all parties, "particularly" the CPSU; the French Party leaders were defensive and ill at ease about criticizing Moscow, explaining that, while deploring the invasion, they would not join in the "anti-Soviet chorus" to which Czechoslovakia was giving rise. The Italian Communists, considerably more acclimated to breathing the heady air of independence of Moscow, went beyond their French comrades by expressing "solidarity" with the Czechoslovak reform movement as well as deploring the intervention. The Italian CP's Directorate on 23 August endorsed the Politburo's position and reaffirmed the party's "deepest disagreement with, and reprobation of, the military intervention." The Italian Party leader, Luigi Longo, had immediately gone to France after the invasion to confer with his counterpart, Waldeck Rochet, reaching a "substantial convergence of opinions," but the unsteadiness of the French posture was clearly evident in comparison with the more forthright Italians.[5]

A brief word is due Fidel Castro's unique assessment of Czechoslovakia. In a speech on 23 August,[6] he engaged in a vigorous rhetorical exercise, leaping from a posture of support for the intervention to positions critical of the Soviets, all the while acutely mindful of the implications of Czechoslovakia for his own regime's situation. Castro brusquely dismissed the official claim that the Warsaw Pact invaders had received an appeal for assistance from the Czechoslovak comrades, remarking with refreshing candor that "besides, there is no lack of fig leaves." The upshot of Castro's assessment was that the socialist camp had the right to prevent Czechoslovakia from falling into the arms of imperialism "one way or another." Though he used the occasion to indulge his own grievances against the Soviets, Castro's endorsement of the invasion of Czechoslovakia marked a reversal in the recent years' trend of cooling Soviet-Cuban relations and an upturn toward closer ties. The new trend would bring Cuba into something approaching a full-fledged member of the Soviet bloc, including membership in the bloc's economic organization.

Castro also used the occasion to call Soviet-U.S. détente into question as "idyllic hopes," and he reflected the affinities among Cuba, North Vietnam, and North Korea when he wondered aloud whether Warsaw Pact divisions would be sent to those countries if they were subjected to "Yankee imperialist attack." These musings expressed Castro's concern to bring home to the Kremlin the implications of Czechoslovakia for the rest of the socialist camp, and thus to remind the Soviets of their global obligations. In doing so, however, he also recognized the implications for the international system itself, and here he reflected the Cuban quandary over how to react to Czechoslovakia in view of its significance for global power politics. He pointed out in his introductory remarks that some of the things he was going to say would "constitute serious risks to our country," explaining this by reference to speculation that Soviet intervention in Czechoslovakia made it proper for the United States to consider a similar intervention in Cuba. Thus Castro, like the Chinese and others assessing the implications of Czechoslovakia, perceived a systemic symmetry underlying Soviet and U.S. power politics. It was a notable development that the logic of this perception impelled the Cubans to move into Moscow's embrace. Geopolitically, if not ideologically, that was no loss to the Chinese, who, in any case, had had their own stormy troubles with Castro. But if a similar logic should similarly impel Peking's neighbors, the North Korean and North Vietnamese regimes, into becoming something approaching protectorates of Moscow as guardian of the socialist camp, then an intolerable menace would be posed to vital Chinese interests. This is the context in which the reactions to Czechoslovakia by Hanoi and Pyongyang must be examined.

NORTH KOREA AND NORTH VIETNAM

In recent years, the Cubans and North Koreans had nourished and celebrated their affinities as Communists and revolutionaries, affinities reflecting their militant and anti-American postures and their fiercely independent stance in the socialist camp. These mutually held positions had also placed them in the forefront of those urging the Soviets and other Communist powers to take greater risks and to undertake stronger commitments on behalf of the Vietnamese revolutionaries. On these matters, and particularly with respect to the interest in a solid socialist camp backing the Vietnamese, the North Koreans and Cubans shared interests with the North Vietnamese that the invasion of Czechoslovakia seemed to these parties to have served.

There were even more fundamental structural factors underlying the affinities between the North Korean and North Vietnamese parties. Each headed a regime controlling the northern half of a divided land, confronted by a hostile anti-Communist regime in the south that was backed by U.S. military and political power. Each had had the experience of direct U.S. military intervention in its respective country, and thus regarded the United States as a fundamental obstacle to the ultimate revolutionary goal of reunification of the country under Communist rule. Consequently, these parties had a strong vested interest in having a solid socialist camp provide a credible deterrent and necessary assistance to shore up vulnerable regimes in a state of hostile confrontation with American power. Moreover, apart from these comparable geopolitical situations, the North Korean and North Vietnamese parties had shown ideological affinities in the first half of the decade as the bipolarizing forces in the Communist movement impelled the radical Asian parties to gravitate toward Peking's leadership in its challenge to Moscow's authority. These ideological affinities had been overlaid and overridden in the mid-1960s by the exigencies of the Vietnam War and sharpened geopolitical imperatives. The Pyongyang-Hanoi axis was thereby tightened, but relations with Peking became severely strained. It was against this background that the Soviet crackdown in Czechoslovakia won North Korean and North Vietnamese endorsement, a position diametically at odds with the Chinese reaction.

The North Korean reaction showed some of the ambiguity of the Cuban one, though more by omission than in the forthright manner of Castro's assessment of Czechoslovakia. Thus, Pyongyang made very clear its revulsion against the Czechoslovak reform movement and accepted the necessity of the Soviet action to squelch such a movement, but North Korean comment on the whole affair all but failed to mention the invasion itself. In Pyongyang's first reaction, the official news agency KCNA on 22 August carried a chronology of Czechoslovak developments since January, which concluded with a single sentence on the invasion, citing the 21 August TASS announcement of the action as having taken place at the request of Prague leaders in order to preserve socialism in that country.[7] Even more striking, Pyongyang's authoritative assessment of the affair, in an editorial in the party paper *Nodong Sinmun* on 23 August,[8] made no direct references to the invasion at all. The editorial, entitled "Historical Lesson of the Czechoslovak Situation," gave a tendentious account of the development of the crisis, observing that the Koreans had been closely watching "the activities of the counterrevolutionary elements in Czechoslovakia from the first day of their coming to the fore." The editorial was intent on drawing what it described as a serious lesson for the international Communist movement on the dangers of revisionism, an approach comporting with Pyongyang's hard-line ideological position and its insistence on the primacy of class struggle. As the editorial put it, the revolutionary essence of Marxism-Leninism lies in its doctrine of class struggle and the dictatorship of the proletariat, a theme characteristic of Pyongyang's Stalinist dogmatism. Stressing the need for joint defense of the socialist camp and for unity against the United States, the editorial emphasized that the socialist camp was "the precious gain which the world working class has won through a century-old bloody struggle." To defend and safeguard the socialist camp, the editorial insisted, was "the sacred duty of each socialist country and all communists." This language seemed tailor-made for justifying the intervention in Czechoslovakia, and clearly that was its import, but why were the Koreans so wary about coming out directly with applause for the invasion?

Both the North Koreans and their friends the Cubans used the Czechoslovak question as an object lesson on the dangers of revisionism, a lesson meant mainly for Moscow's edification. They may have been uncertain at the time as to how fully the Soviets would take the lesson to heart and shed the revisionist proclivities acquired since the Stalin era. Castro also showed his wariness about the implications of the invasion for global politics, especially as a precedent for similar U.S. intervention in presumed American spheres of influence. Pyongyang's wariness was probably of a different nature, reflecting concern not over a precedent for the United States but over a precedent for Soviet intervention in another party's affairs. As much as the Koreans desired to see the Czechoslovak liberalization squelched, and this came through forcefully in their account of the year's developments, they were loath to embrace the act of intervention itself. But this had a certain consistency, for Pyongyang's animus against revisionism had been matched in recent years by its fierce insistence upon independence in the Communist movement. Thus, the invasion of Czechoslovakia was regarded as an act of necessity, but it was not something the Koreans were going to crow about.

For the North Vietnamese, however, the priorities were rather different, notwithstanding the structural affinities noted above.

To be sure, their regard for independence was not to be underrated, and indeed it was an imperative in view of their need to retain the support of both the Soviets and Chinese. But the very exigencies that required this careful balancing act between Moscow and Peking also operated to make the invasion of Czechoslovakia seem not only necessary but even desirable. This was because the invasion served, in Hanoi's eyes, to preserve the integrity of the socialist camp, and this was the very precedent that was welcomed by "the southeastern outpost of the socialist camp," an outpost besieged by American military intervention in Vietnam. For the North Vietnamese it must have been heartening news to learn that the Soviet Union would risk its détente policies in order to preserve the socialist camp, particularly considering that this was a case of defending socialism in one of the marches or outposts of the camp.

Hanoi had remained largely silent in public as the Czechoslovak crisis was brewing, but it promptly publicized the invasion, publishing the text of the TASS announcement along with a preface which amounted to a clear endorsement. Thus, Hanoi's preface explained that increasing "anti-socialist" activities by "counterrevolutionary forces" in Czechoslovakia since January had seriously jeopardized the socialist system in that country, "a member of the socialist camp." The recent intensification of hostile activities by those counterrevolutionary forces had "compelled" the "staunch" members of the Prague leadership to request the assistance of the Warsaw Pact allies. The Pact's forces had entered Czechoslovakia with the "noble aim" of preserving socialism there, the preface said. Both the North Vietnamese party organ *Nhan Dan* and the military paper *Quan Doi Nhan Dan* front-paged the TASS text along with Hanoi's preface; the grateful Soviets as promptly took note of this in *Pravda*, which quoted the preface in full.[9] Moscow was probably even more pleased when the Hanoi press, a few days later, carried the full text of a lengthy *Pravda* editorial article of 22 August that set out an authoritative Soviet interpretation of the Czechoslovak crisis and its meaning.[10]

Hanoi had thus promptly and clearly signaled its endorsement of the intervention, and had received grateful feedback from Moscow. As it happened, there was a major occasion two weeks after the invasion for Hanoi to issue a general policy statement, thereby affording an opportunity for the North Vietnamese to interpret Czechoslovakia within a basic strategic and ideological framework. The occasion was the anniversary of the founding of the Democratic Republic of Vietnam (DRV), on 2 September, and the policy statement was delivered by Premier Pham Van Dong at a Hanoi meeting marking the event.[11] As prefigured in Hanoi's preface to its publication of the TASS announcement of the invasion, in which it pointedly identified Czechoslovakia as a member of the socialist camp, Pham Van Dong's speech not only made much of the present importance of the socialist camp but also put the Czechoslovak question squarely within the context of the defense of the camp. Significantly, the premier preceded his remarks on Czechoslovakia with a discussion of aid to Vietnam in terms of defense of the socialist camp, declaring that the Vietnamese were deeply aware that their victories could not be separated from "the very great and valuable support and assistance of the socialist countries." Underscoring Hanoi's sense of the mutual internationalist obligations arising from the Vietnam war and the DRV's membership in the socialist camp, he expresssed gratification that standing alongside the Vietnamese in the anti-American struggle were "over one billion people of the socialist countries [who] are fulfilling their internationalist obligations to our people—the fighters standing firm on the front-line to defend the socialist camp who are resolved to accomplish their internationalist duty."

Pham Van Dong's linkage of the defense of the DRV with the preservation of socialism in Czechoslovakia came in the context of a paean of the Communist alliance system. Echoing the North Koreans, he expressed the view that "the union of all socialist countries in a great family to form the world socialist system is the greatest achievement of the revolutionary struggle." That was why, he continued, "our people are filled with great joy and enthusiasm when socialism records a new victory and with deep concern when socialism faces a threat in any part of the world." And here he plunged directly into the Czechoslovak question as the source of just such a concern over a threat to socialism. Because of the serious danger threatening socialism in Czechoslovakia, he explained, the Warsaw Pact armies were compelled to intervene to preserve socialism there and to maintain that country in the Pact and the socialist camp.

The Vietnamese were fully aware of the bitterly hostile reaction by the Chinese to the invasion of Czechoslovakia but, once again, as in the crucial turning point in the middle of the decade, the role of the socialist camp had become a touchstone marking fundamental divergences of policy and priorities. As Peking's own denunciations of the Dubcek leadership indicated, the Chinese shared with Hanoi and Pyongyang a deep revulsion against the Czechoslovak liberalization movement that the Soviet invasion sought to squelch. There was also a congruence of interest among these Asian Communist regimes in the disruptive effect that the invasion might be expected to have on Soviet-U.S. détente. These were the factors that underlay the ideological affinities of the Asian Communist parties in the early 1960s, but the major United States involvement in Vietnam had lent new urgency to fundamental issues of socialist camp solidarity and international Communist relations, and the post-Khrushchev Kremlin leadership's line of united action on Vietnam had made these issues all the more momentous for the Chinese. Now, in 1968, the Czechoslovak crisis, and especially the reactions to the Soviet intervention, had invested those issues with even sharper significance. This is why the Chinese response must be viewed against the background of these reactions, particularly Hanoi's, and why Czechoslovakia had such a profoundly transforming effect, even though, paradoxically, its impact seemed to have confirmed rather than challenged the international system. The Chinese decided that the situation demanded their own undertaking of such a challenge to the existing system. The logic of that decision led to bloodshed along the Sino-Soviet border and to a dramatic new turning point in Sino-American relations.

EVOLUTION OF PEKING'S RESPONSE

Before August, the Chinese had had little to say publicly on the Czechoslovak developments, alluding only vaguely to a situation in which "modern revisionism with the Soviet revisionist renegade clique at its center" was "disintegrating." On 10 August, a week after the Bratislava meeting between the Prague leadership and the leaders of the five Warsaw Pact countries that later invaded Czechoslovakia, the Chinese news agency NCNA belatedly carried an extensive summary of a 24 July Albanian *Zeri I Popullit* interpretation of the Czechoslovak crisis.[12] The timing of the NCNA release indicated that it was prompted by Peking's assessment of the Bratislava conference, rather than the merits of the Albanian interpretation as such. Peking's failure to offer an assessment of the situation in its own name suggested that it viewed the matter as still quite fluid and uncertain of outcome; in the meantime the use of the Albanian article would serve to discredit Moscow's motives, whatever the ultimate outcome. According to the Albanian interpretation, the Czechoslovak crisis grew out of a plot organized by Moscow that got out of control, but the Soviet leaders were not likely to order an invasion because that would torpedo the International Communist Conference scheduled for November. The article predicted that the Soviets would be going to their Canossa in order to paper over their differences with the Prague leadership; and indeed, on the same day that NCNA belatedly publicized the 24 July article, *Zeri I Popullit* carried an editorial article gleefully seizing upon the compromise Bratislava statement as having confirmed that prediction. The new article explained the retreat from the ultimatum-like terms of a menacing letter in July to Prague from the Warsaw Five as due (1) to unwillingness to spoil Soviet-U.S. relations by military intervention in Czechoslovakia, and (2) to fear of being isolated in the international Communist movement at a time of preparations for the November conference.[13]

Once the invasion had invalidated the Albanian prediction, the Chinese immediately came out with an assessment in their own name, though here also there was a significant evolution of line as reactions to the event as well as the event itself became a factor. The first Chinese reaction came in the typical form of a *People's Daily* "Commentator" article on 23 August coupled with an NCNA news report of the invasion portraying it as the outcome of acute conflicts within the Soviet bloc, of an acute struggle between the United States and the Soviet Union for control of East Europe, and, finally, of the "collaboration" between the United States and the Soviet Union to "redivide" the world. The Commentator article dismissed the Soviet claim to be defending the socialist community as signifying in fact an attempt to maintain a colonial empire. The article introduced a highly significant new charge to Peking's anti-Soviet indictment, declaring that the "Soviet revisionist renegade clique has long ago degenerated into a gang of social-imperialists."[14] This charge of "social-imperialism" was to be the keynote of a new stage of Chinese hostility toward the Soviets.

As it happened, there were two diplomatic occasions in the immediate wake of Czechoslovakia that were tailor-made for the Chinese to deliver their message to the international community, and on these occasions they signaled the two major lines of their response to Czechoslovakia. The first, on the same day as the 23 August "Commentator" article, was the Romanian National Day; Chou En-lai used this occasion to deliver a scathing denunciation of the Soviet intervention in Czechoslovakia, and, in the process, he foreshadowed the opening of a new and active East Europe policy by Peking. The immediate purpose of this policy line was to deter the Soviets from further armed ventures against maverick Communist regimes in East Europe. The longer-term aspect of this policy was the evolution of what Moscow would call Peking's differentiated line toward the Soviet bloc, a line aimed at fostering vested interests in independence of Moscow and to reward attempts at resisting Soviet control over East Europe.

Impact of Czechoslovakia, 1968

The second occasion, already noted in connection with Hanoi's reaction to Czechoslovakia, was the DRV's anniversary on 2 September. Again Chou delivered a major attack on the Soviets, in a speech at the DRV embassy, and here Peking's reaction to the invasion can be seen as responsive to others' reactions as well—that is, it was reaction to others' reactions as well as to the event itself, and Peking's concern here was essentially with the implications of Czechoslovakia for the Asian Communist movement. In 1965 the Vietnam question, particularly in its implications for Soviet influence in Asia (and hence for Chinese independence), had a profoundly transforming effect on alliance relationships, in effect suspending the Sino-Soviet alliance. Now the implications of Czechoslovakia would carry that process still further toward a profoundly transforming effect on adversary relationships. This process, following along the second major line of Peking's response to Czechoslovakia, was significantly foreshadowed in Chou's 2 September speech at the DRV embassy. But to distinguish these two lines it is necessary to examine the evolution of Peking's approach in the short period bounded by the two major Chou speeches.

In his 23 August speech at the Romanian embassy, Chou quickly launched into a denunciation of the invasion of Czechoslovakia as "the most barefaced and most typical specimen of fascist power politics" practiced by the Soviets against their own allies. He declared that the Soviet leadership had long since "degenerated into social-imperialism and social-fascism." Significantly, Chou placed Czechoslovakia in the context of Soviet-American global relations, arguing that the United States acquiescence in the invasion of Czechoslovakia meant that the Soviets were not able to oppose U.S. control of South Vietnam. He then generalized a lesson from this situation: "That a big nation should have so willfully trampled a small nation underfoot serves as a most profound lesson for those harboring illusions about U.S. imperialism and Soviet revisionism." Chou concluded his long discourse on the meaning of Czechoslovakia by declaring flatly that Romania itself was now facing the danger of intervention, and he assured the Romanians that the Chinese people "support you."[15]

Peking's moves to signal support for the Romanians and other East Europeans feeling menaced by Moscow could hardly have been more than declaratory at this early date. What was significant was that Chou's warning of a threat to Romania signaled the opening of an East Europe policy through which Peking hoped to secure leverage against Soviet activity in that area. Perhaps even more significant, this move, together with a comparable step by the United States, represented the beginning of a potentially new factor in the international system: parallel moves by the United States and the PRC aimed at checking Soviet hegemonistic thrusts. This parallelism was illustrated when, a week after Chou's warning of further Soviet moves in East Europe, President Lyndon Johnson took note of what he delicately called "rumors" that the Soviet action in Czechoslovakia "might be repeated elsewhere in the days ahead in Eastern Europe" and warned that the "dogs of war" must not be unleashed. These were of course parallel rather than coordinated moves, but this process would arrive at the point a few years later where the United States and the PRC would join in a historic declaration (the February 1972 Shanghai communiqué) pledging their efforts to oppose any country's or bloc's hegemony in East Asia.

New comment from Peking on Moscow's practice of power politics was elicited when, on 27 August, a Soviet accord with the Dubcek leadership was reached, thereby temporarily shelving Moscow's effort since the invasion to replace Dubcek with elements responsive to the Kremlin. Another *People's Daily* 'Commentator" article, "Deal Made at Bayonet Point,"[16] explained the reversal of Moscow's earlier authoritative damnation of Dubcek as the result of pressure by the United States on the Soviets and of "intensified U.S.-Soviet counterrevolutionary global collusion." Thus, the talks in Moscow with the Dubcek leadership were "a component part" of this global collaboration, in which the United States gave tacit consent to the invasion of Czechoslovakia while bringing pressure on the Soviets to "attach first importance to the overall state of U.S.-Soviet relations." Czechoslovakia, in Peking's interpretation, was a chip in global bargaining by the two big powers, and maintenance of the existing international order was a controlling aim to which issues like Czechoslovakia and Vietnam were subordinated by Washington and Moscow alike.

To ensure that the meaning of the significant new charge of "social-imperialism" was understood, *People's Daily* on 30 August called attention to its use of the term a week earlier and quoted Lenin's definition of the term as meaning imperialism flying the flag of socialism, or "the growth of opportunism into imperialism." In its explanatory note, *People's Daily* said the invasion of Czechoslovakia was "the most typical and conspicuous exposure of [the Soviet leaders'] ugly features of social-imperialism." NCNA disseminated this note worldwide and it was also published in the *Peking Review*[17] in a box headed "Glossary." As in other instances of important signals marking a new phase in Sino-Soviet relations, a new formula codified and concentrated a range of meanings and implications, thus providing a methodological key for understanding complex processes.

The next diplomatic occasion, the DRV's anniversary on 2 September, produced a speech by Chou En-lai so rich in its implications that its significance can hardly be overestimated. (It also provides an excellent methodological case study.) The key contextual factors in interpreting its meaning were the occasion itself, in which the Vietnamese were the primary audience for the message being delivered, and the impact of Czechoslovakia, particularly Hanoi's reaction in the nearly two weeks since the event. Also of methodological significance were the striking contrasts between the policy statements delivered on this occasion by the Chinese and North Vietnamese premiers, contrasts that brought into sharper relief the divergent assumptions and priorities underlying their respective policy lines.

Closely related to Peking's recent introduction of the charge of "social-imperialism" was Chou's straight-out assertion on the DRV anniversary that the Soviet leadership had "long since completely destroyed the socialist camp which once existed." Though in each of these charges the Chinese claimed that the condition being ascribed was of long standing, the significance in context lay in the *act* of saying so. What was striking here was not only that Chou chose to pronounce obsequies on the socialist camp before this particular audience, but that this marked the first time the Chinese had made fully explicit their renunciation of the Communist alliance system. True, Lin Piao's major statement of strategy in his September 1965 tract on people's war had failed to invoke the socialist camp, a telling omission at a time when rising hostilities in Vietnam had sharply posed the question of the alliance's role; but the rejection of the socialist camp had been left implicit for three years. In fact, even an authoritative Chinese policy statement of 11 November 1965 effectively closing the debate on strategy referred, albeit in an offhand and perfunctory manner, to the socialist camp, and the term itself only gradually withered away from Peking's lexicon.[18] The three-year delay in making an explicit statement on the matter indicated its sensitivity; this reflected the importance attached to the socialist camp during that time by the North Vietnamese and North Koreans, and perhaps also reflected reluctance on the part of influential elements in the Chinese leadership to close the door finally on something that had once been the cornerstone of mutual security. Whatever the previous inhibitions, Chou's unequivocal dismissal of the socialist camp on this occasion represented an important decision that signaled an important move in international Communist politics.

That it was others' reaction to Czechoslovakia rather than the event as such that chiefly motivated Peking's move was indicated by the textual context of Chou's remark on the socialist camp. Inasmuch as Hanoi had endorsed the invasion as a necessary measure to preserve the socialist camp, Chou's explicit renunciation of that concept—he also quoted Moscow's own invocation of "the socialist community"[19]—seemed intended mainly for Vietnamese ears, as the occasion itself indicated. It was gratuitous, and significantly so, for Chou to have resurrected—if only for public burial—the concept of a socialist camp; this was the more obvious inasmuch as in the same breath he directly quoted Moscow's reference to the socialist community, a term that would have sufficed for denouncing the invasion as such. But he was especially commenting on Hanoi's favorable reaction to the event; moreover, Chou seemed to have been addressing the Vietnamese in particular when he went on to remark bitterly that to describe "this barbarous fascist aggression as Marxist-Leninist and proletarian internationalist aid is nothing but a flagrant betrayal of Marxism-Leninism." And he really brought the point home to his DRV hosts when, in placing the Czechoslovak question in the framework of U.S.-Soviet global relations, he raised the Vietnam question as another case in point. He argued that, since the United States had recognized Czechoslovakia and the rest of East Europe as within the Soviet sphere of influence, "the condition in return" was naturally that the Soviets should recognize the Middle East, Vietnam, and the rest of Southeast Asia as falling within the U.S. sphere. Thus, after Czechoslovakia, the United States would "definitely demand a higher price" on the Vietnam question, while the Soviets would serve the U.S. interest "all the more obsequiously." And here came the main message: "It is high time all those who cherish illusions about Soviet revisionism and U.S. imperialism woke up!"[20]

Could there have been any doubt *in this context* about who "those" were who were cherishing illusions about the Soviets? The implication was that Hanoi's illusion consisted in interpreting Soviet intervention in Czechoslovakia as a precedent for Soviet defense of the integrity of the socialist camp wherever it was threatened. In the Chinese view, buttressed by their own experience of Soviet priorities in managing the affairs of the socialist camp, the Vietnam question, like Czechoslovakia, was another bargaining chip in Soviet-U.S. global politics. The Chinese perceived Moscow's behavior as joining with the United States in seeking to preserve a bipolar international system, and in this view there was a systemic symmetry between Soviet action in East Europe and United States action in Vietnam, among other areas.

It should be noted that, on both the Romanian and DRV diplomatic occasions, Chou addressed those harboring "illusions" about the United States and the Soviet Union; however, there were two differences in his treatment of this theme having particular analytical significance. The first difference was the sharply heightened polemical tone, the cutting edge of his statement delivered on the second occasion. This expressed Peking's exasperation and dismay over Hanoi's strong endorsement of the Soviet invasion, prompting Chou to deliver a stern lecture to the Vietnamese comrades. That is to say, this in effect reflected a difference between Peking's reaction to Czechoslovakia as such and its reaction to Hanoi's reaction to the event. The second significant difference, one that signaled a new direction that would have immense implications, was the reversal of the order of "U.S. imperialism" and "Soviet revisionism," Chou having put the latter first in his statement on the second occasion. This reversal represented a striking departure from established practice, and, from a methodological standpoint, it can be interpreted as a type of early warning signal of a development that would reorder the basic adversary relationships shaping the international system. This development was itself subject to twists and turns before issuing in concrete results, and it was by no means irreversible; indeed, it was destined not only to be the product of a complex process of signaling and feedback among the major powers, but it also was to become a factor in another major power struggle within the Chinese leadership. At this early incipient stage, however, this signal already reflected the profound implications of Peking's assessment of the meaning of Czechoslovakia, particularly Peking's concern over the implications for international Communist relations. As events would soon show dramatically, the Chinese were determined to invest concrete meaning in their charge of Soviet social-imperialism, and in so doing they put teeth into their words, as it were. Thus, the operative meaning of Peking's statements about social-imperialism and the socialist camp was not to describe or report a state of affairs that had "long since" existed; rather it lay in the act of making these charges, and all that electing to do so implied. The Chinese were *signaling* something to come, not reporting what had already taken place, and they thereupon proceeded to act in significantly new and demonstrative ways. In particular, as will be seen, the Chinese acted so as to demonstrate that the Soviets indeed behaved like imperialists in Asia and that the Kremlin leaders were the "new tsars" pursuing old imperialist aims.

Peking's global-political interpretation of Czechoslovakia was given further definition and application shortly after Chou's two major statements in speeches by Foreign Minister Ch'en I. Again a diplomatic occasion conveniently played into Peking's hand, this time the anniversary of the Democratic People's Republic of (North) Korea (DPRK), a regime whose stand on Czechoslovakia must also have perturbed the Chinese. Speaking at a DPRK embassy reception on 9 September,[21] Ch'en undertook the job (which required doing) of distinguishing between the Soviet intervention in Czechoslovakia that year and the 1956 invasion of Hungary, which Peking had endorsed. Ch'en's denial of the analogy turned on his explanation that, in 1956, "Khrushchev revisionism was only beginning to raise its head" in the Soviet Union, which had not yet entered on the road of collaboration and détente with the United States. This explanation accorded with Peking's standard dating of its troubles with the Kremlin as beginning with the Soviet Party's 20th Congress in February 1956, only a few months before the Hungarian uprising. This also implicitly makes sense of Peking's notable efforts in the first few years after Hungary to make the socialist camp a central instrumentality in behalf of Chinese interests. But now, Ch'en argued, Moscow's invocation of the socialist community was "simply a synonym" for a Soviet sphere of influence, a component part of the framework of Soviet-U.S. global relations. To show that the international system as confirmed by Czechoslovakia adversely affected Pyongyang's own interests, Ch'en discoursed on the Soviet role as an alleged accomplice of the United States in Northeast Asia, including Moscow's tacit recognition of South Korea as being within the U.S. sphere of influence. Here again the Chinese were depicting a systemic symmetry in the two superpowers' behavior in a bipolar system. This was further spelled out by the Chinese foreign minister in a speech on 19 September welcoming his South Yemeni counterpart.[22] In the 1962 Caribbean crisis, Ch'en explained, the Soviets recognized Cuba as lying within the U.S. domain, and

Czechoslovakia in turn demonstrated the pattern in which each of the two superpowers "can allow the other to have a free hand in doing whatever evil it likes in its sphere of influence." Bringing the lesson home to another audience, this time to the Arabs, Ch'en warned that the Soviet invasion of Czechoslovakia had strengthened the American and Israeli position in the Middle East.

FOUR

East Europe and Sino-Soviet Border Tension

In addition to the charges of Soviet social-imperialism and destruction of the socialist camp, Peking at this time introduced still another charge that referred literally to longstanding conditions but whose operative meaning lay in signaling a new phase in the Sino-Soviet relationship. In this case, the subject was of the most sensitive and explosive nature, the massive Soviet military presence along China's border and Peking's territorial grievances against the Soviet Union. If ever there was a way of giving the Sino-Soviet conflict the character of a fundamental geopolitical confrontation, the Chinese surely must have elected the most sure-fire way by choosing at this time to make an issue out of border tensions.

Peking's initial move in this direction came on 16 September 1968 when, for the first time, it issued a public protest against Soviet air intrusions into Chinese territory. A Foreign Ministry note[1] charged that Soviet military aircraft had flown 29 sorties over Chinese airspace in the northeast province of Heilungkiang between 9 and 29 August. Though it claimed that such violations had been occurring persistently in "recent years," including 119 intrusions in the past year, the note pointed out that it was "rare that within a short space of 21 days the Soviet side should have committed such concentrated, frequent, barefaced, and flagrant military provocations over China's airspace in one area." The Chinese did not leave it implicit that the alleged concentration of violations occurred around the time of the invasion of Czechoslovakia: the note argued that these intrusions were "organized and planned . . . in support of [the Soviet government's] atrocities of aggression against Czechoslovakia and in pursuit of its global strategy of allying with the United States against China and attempting to dominate the world in collusion with the United States." The note concluded with a "stern warning" that "China's territorial integrity and sovereignty absolutely brook no violation."

There were several aspects of the timing and content of this protest pertinent to its operative meaning. First, of course, was the very fact that the Chinese chose to break their silence on a matter of long standing that might have been expected to figure, but had not, in the demonstrative moves of recent years to give expression to the deterioration of Sino-Soviet relations. Furthermore, the note indicated that this decision to air the sensitive border troubles was related to Czechoslovakia. In fact, the Chinese seemed to have strained somewhat to make that connection for, by their own reckoning, 20 of the 29 alleged sorties in August took place by 12 August, more than a week before the Warsaw Pact troops moved into Czechoslovakia. The alleged violations did not seem to have been all that serious; the Chinese claimed that the Soviet aircraft penetrated 3 kilometers at the farthest and covered 5 kilometers at the longest. The real significance of the protest, as its timing would indicate, must be sought in the context of Peking's reaction to Czechoslovakia as this was evolving at the time the note was released.

A few days later, on 21 September, Peking expanded significantly on the protest note's charges of territorial violations. A *People's Daily* article said to have been written by a soldier in *north* China delivered a warning to (in this order) "the Soviet revisionists, the U.S. imperialists, and reactionaries of various countries" regarding their alleged design to set up a ring of encirclement around China.[2] The article claimed that the recent Soviet air intrusions, which had aggravated tensions on the border, were part of organized and planned actions in coordination with United States global strategy to encircle China. It went on to add that, though this was "the first aim" of the United States and the Soviet Union, they also sought to control China's neighbors under the cover of a strategic encirclement of China. This was a warning intended for the fraternal Vietnamese and Koreans, who had failed to share Peking's perception of Soviet intervention in Czechoslovakia as the behavior of an imperialist power.

Also during this period, and closely related to the issue of border tensions now being surfaced, the Chinese developed a vigorous and highly demonstrative show of support for Albania's defiance of the Soviets, a particularly significant defiance in the context of widespread speculation that the Soviets might be inclined to extend their recent exercise of force in East Europe to regain control of the maverick Balkan Communist regimes. Whatever might have been Peking's role in Tirana's decision, Albania's announcement of formal withdrawal from the Warsaw Pact in the wake of the invasion of Czechoslovakia admirably served Peking's moves to undercut Moscow's invocation of the Communist alliance system to justify Soviet behavior. What was particularly significant was not only the authoritative character of Peking's support for Albania's defiance, but also the way the Chinese interwove this campaign with their moves to make an issue out of the inflammatory Sino-Soviet border tensions. By the skillful use of this linkage the Chinese were able to put their emerging East Europe policy in the service of their effort to show the Soviet Union to be an imperialist power seeking hegemony in Asia.

The sequence of events is revealing. On 12 September, Premier Shehu proposed to Albania's parliament the law declaring that country's formal withdrawal from the Warsaw Pact. In explanation, Shehu said that to remain in the Pact following the invasion of Czechoslovakia would be tantamount to participating in an anti-socialist alliance and abetting U.S.-Soviet collaboration for world domination. He pointedly reminded the Soviets that the Albanians "are not alone—they have numerous faithful friends," in the first place "great People's China with a population of over 700 million." On the next day, the proposal was enacted into law by the Albanian parliament and, on the 15th, Peking reported this action. On the 16th, Peking issued the protest of Soviet intrusions into Chinese airspace. On the next day, in addition to carrying a lengthy account of Shehu's speech, including his reference to Chinese support, Peking issued a message signed by Mao, Lin Piao, and Chou En-lai expressing "firm support" for Albania's withdrawal from the Warsaw Pact and asserting that the Chinese "will at all times and under any circumstances unswervingly stand on the side of the fraternal Albanian people." Apart from the level of the signators, the sense that an important change was taking place in a wider context was conveyed in the message's declaration that a "new historic stage" of opposing the United States and the Soviet Union had begun.[3]

Still another signal of the importance Peking was attaching to these developments was the appearance on 20 September of editorials in both the *People's Daily* and *Liberation Army Daily*[4] hailing the Albanian move as "an important contribution to the international communist movement" and again proclaiming a new stage of opposition to the United States and the Soviet Union. After analyzing Moscow's use of the Pact as a means for pursuing global collaboration with the United States, the *People's Daily* editorial cited Czechoslovakia as evidence that a country's membership in the Pact provided no guarantee of its sovereignty and territorial integrity. What would stop the Soviets, the editorial asked, "from mustering a number of member countries to do the same against still another member country tomorrow?" Thus, as in Chou's speech at the Romanian embassy shortly after the invasion, the Chinese were raising the question—which had been raised also by President Johnson—of a possible Soviet strike against the Balkan mavericks. Whether or not Peking was inviting Romania to follow Albania's example in withdrawing from the Pact, a rather fanciful hope, it was clear that the Chinese were bent on encouraging defiance of the Soviets and on doing what they could to thwart Moscow's efforts to consolidate its hold over other Communist countries. Put another way, the Chinese were intent on demonstrating that the Soviets were indeed social-imperialists rather than proletarian-internationalist defenders of the socialist camp. After all, as Chou had forcefully lectured the North Vietnamese, the Soviets had long since destroyed the socialist camp, and the Chinese were determined to prevent them from putting that camp together again.

The sequence of these events was suggestive enough, but subsequent events, still continuing in a Sino-Albanian framework, underscored even more clearly the link between Peking's interest in East Europe and the Sino-Soviet border situation. The theme of Sino-Albanian mutual security was given even more direct expression by the arrival in China on 28 September of the Albanian Defense Minister, Beqir Balluku, a Politburo member and Deputy Premier in addition to his defense portfolio. A welcoming editorial carried by both the *People's Daily* and the *Liberation Army Daily* on the 29th[5] raised the question of a Soviet threat to the Balkans, saying that after Czechoslovakia the Soviets had amassed a large number of troops in Bulgaria in a threat to security in the Balkan region. The editorial closed with a declaration that the Chinese and Albanians would "always fight shoulder to shoulder," but there was no reference to a threat to Chinese security. However, later that day, at a banquet honoring the Albanian delegation, Chou En-lai explicitly linked the question of Albanian security and tensions along the Sino-Soviet border that had been the subject of the 16 September protest. Chou said that, while posing a threat to security in the Balkans by massing troops in Bulgaria, "Soviet social-imperialism is also stepping up armed provocations against China. In coordination with U.S. imperialism it is energetically forming a ring of encirclement against China by stationing massive troops along the Sino-Soviet and Sino-Mongolian borders and, at the same time, it is constantly creating border tensions by ever more frequently sending planes to violate China's airspace."[6] Chou thus put into one explosive package the charge of Soviet overflights introduced earlier in the month, the recent charge of Soviet troop concentrations in Bulgaria menacing the Balkans, and a revival of the charge of Soviet troops concentrated along the Chinese border that had been notably dormant for twenty months of the most serious deterioration in Sino-Soviet relations. The charge of Soviet troop concentrations along the border had last been aired at the elite level in January 1967 by Yeh Chien-ying at a rally for an Albanian delegation headed by the same Balluku. Yeh had claimed that Moscow was "shifting its military strategy to the East" and was "stepping up its disposition of troops against China." The Soviet military build-up opposite China had been underway for a year or so by the time Yeh spoke, and border troubles dated back to the beginning of the decade; yet this highly inflammatory subject was thereafter avoided publicly by Peking, even during periods in which Sino-Soviet tensions were being acrimoniously aired on just about every other grievance. Against this background, Peking's revival of the issue of border tensions in the wake of Czechoslovakia was ominous.

The theme of mutual security was made more pointed in speeches at a Peking rally on 4 October honoring the Albanian delegation. The Chinese People's Liberation Army (PLA) Chief of Staff, Huang Yung-sheng, spoke at length on the implications of Czechoslovakia, equating Soviet relations with East Europe and U.S. relations with Latin America. He repeated the linkage Chou had made by charging that Moscow had massed troops in Bulgaria, "thus posing a grave threat" to Balkan security, and had reinforced its troops along the Chinese border and "thereby ... intensified its armed provocations against China." After citing Mao's message to the Albanians the previous month as warning the United States and the Soviet Union against attacking Albania, Huang offered direct assurance to the Albanians that the PLA was determined to "fight shoulder to shoulder" with their army. In his speech, Balluku rhapsodized over China's role as the motherland of the world proletariat and rejoiced that "the world standard-bearer of Marxism-Leninism" was Albania's "closest ally, friend, brother, and comrade-in-arms."[7]

Apart from raising the question of a possible Soviet thrust into the Balkans, the Chinese during this period were also raising the specter of a possible invasion of their own country. Thus Chou, speaking on 30 September on the eve of the PRC's National Day,[8] called for heightened vigilance and readiness "to smash any invasion launched by U.S. imperialism, Soviet revisionism, and their lackeys, whether individually or collectively." The phrase "individually or collectively," a new formulation, admitted of the possibility of a Soviet invasion alone; the most direct evocation of a Soviet invasion threat was voiced a few days later during the Albanian delegation's visit to Sinkiang, the highly sensitive border region where China's nuclear test facilities were located and a region that in recent years had been off-limits for visiting delegations. Speaking at a banquet for the Albanians on 7 October, Wang En-mao, the long-time military boss of the region, declared that the PLA units in Sinkiang were strengthening their preparedness against war and were "consolidating frontier defense." In this context, he raised the possibility of the Soviets' daring to attack.[9]

In another sign at this time of Peking's burgeoning East Europe policy, the Chinese in mid-September inaugurated broadcasts to Czechoslovakia, Poland, and Romania, three countries where they evidently hoped to find receptive anti-Soviet audiences.[10] A dramatic expression of Peking's growing East European activity came later that year when Huang Yung-sheng led a delegation to Albania, marking the first foreign trip by a member of the Chinese elite since June 1967.[11] In a striking if not provocative show of interest in Romania's independence, Huang made a stopover en route at Bucharest at the very time that the Warsaw Pact members' chiefs of staff were meeting there. The Romanian Chief of Staff took time off to see his Chinese counterpart, a gesture that must have reddened the cheeks of the Soviet opposite number. While in Albania, Huang further underscored the theme of Sino-Albanian mutual security that had been featured in the Albanian Defense Minister's visit to China and, in a speech at a Tirana rally on 2 December, he again juxtaposed the charge of Soviet troops concentrated in Bulgaria threatening Balkan security with the allegation of Soviet provocations along the Chinese border.[12]

In the meantime, Moscow must have been eyeing the growing

Chinese interest in East Europe with watchful concern, and Peking's decision to raise the border issue at the same time could only have been viewed with acute apprehension. As subsequent events would further confirm, the Soviets were loath to become engaged in a territorial dispute with the Chinese, or indeed with other countries having latent territorial or irredentist grievances (the Soviets regarding the postwar territorial settlement as virtually final). Moscow remained silent on Peking's 16 September protest of air intrusions until 31 October, when a Soviet Foreign Ministry note[13] explained that a comprehensive investigation failed to confirm any of the Chinese allegations. The note provided a glimpse into existing border tensions by remarking that on "several occasions" Chinese border organs had accused the Soviets of intrusions. According to the note, there had been only one intrusion, in December 1967, when a Soviet helicopter accidentally strayed into Chinese airspace, prompting an apology by Soviet border officials. Alluding to the recent Chinese campaign portraying a Soviet threat to the PRC, the note attributed "all these inventions and outcries" to an effort by Peking to "mask the anti-socialist essence" of its position on Czechoslovakia. Interestingly, however, neither in the note nor in other commentary did Moscow directly acknowledge or mention the charge of Soviet troop concentrations along the Sino-Soviet border, though Moscow had referred to the charge of Soviet troops massed in Bulgaria.

This carefully selective treatment of Peking's charges comported with Moscow's concern during the most tumultuous phases of the Cultural Revolution to project an image of restraint and responsibility in dealing with border tensions. In a follow-up to the Foreign Ministry note, *Izvestiya* on 2 November carried an interview with a Soviet deputy chief of staff describing strict precautionary measures designed to prevent inadvertent flights into neighboring countries. Putting his finger on the crucial distinction, he acknowledged that unforeseen events can happen in cases of bad weather or equipment failure, "but if we have good-neighborly relations these can be resolved easily by border organs."[14] The Soviet official thereby indicated not only that border issues were essentially political but also that Moscow hoped the border question could be left dormant and treated only as a matter to be handled by local border organs. This soon proved to be an empty hope; Peking was now determined to press the issue.

While shying away from the explosive border issue, Moscow reacted more openly to Peking's emerging East Europe policy. Notably, an article in the Soviet foreign affairs weekly *New Times*[15] gloomily cited Western press reports saying the Chinese were expanding their military presence in Albania, whose "strategic position as a gateway" to East Europe allegedly served Peking's anti-Soviet interests. The article cited reports that the Tirana airport was closed from 25 September to 2 October—a period overlapping the Albanian Defense Minister's visit to China—because Chinese specialists were engaged in secret work there. As for Huang's visit, the article cited reports that the Chinese delegation signed a secret military agreement permitting the Chinese to station troops and build military bases on the Adriatic coast. Indicating that the Soviets were carefully monitoring Chinese military activities in Albania, the article issued a scarcely veiled warning against going too far. Quoting *Le Monde* as saying there were indications that Peking planned to install offensive missiles, the article observed that this would "seriously impair the security of Albania." It was as if the Soviets feared Peking would attempt in Albania what they themselves had tried to do in Cuba.[16]

FIVE

China, Vietnam, and the United States

Within the brief space of a few weeks after the invasion of Czechoslovakia, Peking had taken several interrelated initiatives that together signaled what it called a "historic new stage" in its effort to cope with the realities of a bipolar system. As both the timing and the substance of these initiatives indicated, the new stage of policy was responsive primarily to Soviet behavior, the type of behavior the Chinese were now denouncing as "social-imperialist." These initiatives, including the decision to declare that the socialist camp no longer existed and to make an issue again of border tensions, indicated not only that Czechoslovakia provided a sharp stimulus to rethinking the international situation but also that the decision-making environment in Peking was becoming propitious for new policy departures in the light of that rethinking. These departures from the isolationist phase of Chinese policy that attended the upheavals of the Cultural Revolution represented a part of a broader trend away from the basic spirit and style of the Cultural Revolution. Moreover, this trend posed increasingly sharp questions about Chinese leadership politics, for the ascendance of elements in the leadership promoting the new initiatives perforce presented a corresponding challenge to others whose status and power were invested in the previous policies. As a rough first approximation, it can be noted that Chou En-lai served as the principal spokesman of the initiatives in question, and that Lin Piao had been most closely associated with the previous, isolationist phase marking the high tide of the Cultural Revolution. These figures provide key points of reference for identifying broader trends in the convulsive competition between those forces having vested interests in the Cultural Revolution and those who now saw an opportunity to enter a period of reconstruction, if not in fact a restoration of the status quo ante.

The immediate post-Czechoslovak period coincided with the completion of the process of forming "revolutionary committees" in all the Chinese provinces, thereby rounding out one major phase of the Cultural Revolution. These committees served as the provincial and local power organs after the dismantling of the party apparatus beginning with the Shanghai "January storm" of power seizures twenty months earlier; they comprised activists who came to the fore during the Cultural Revolution, military commanders who had filled the vacuum of authority created by the party's dismantling, and veteran cadres permitted to continue to function as a third element in this "three-way alliance." The final two revolutionary committees at the provincial level, for the border regions of Sinkiang and Tibet, were established on 5 September 1968.

Two days later, a joint editorial in *People's Daily* and *Liberation Army Daily* saluting the "extremely magnificent spectacle" presented by the completion of this process stressed, significantly, the need now for unified leadership and coordinated actions.[1] The editorial contained a generalized call for strengthening border defense against any (unnamed) "enemy" who dared to invade. Separate reports on the formation of the committees in Sinkiang and Tibet took more pointed note of those regions' sensitive location. NCNA's report on Sinkiang[2] stressed that the region occupied "an extremely important strategic position" as an outpost in the struggle against Soviet revisionism, U.S. imperialism, and the Indian reactionaries; and regional leader Saifuddin, a Uighur, was quoted as warning of Soviet plotting to subvert the region. An NCNA report on the Tibet revolutionary committee[3] was less urgent in tone and transposed the United States and the Soviet

Union in the order of enemies. At any rate, Peking now had firmer control over these vulnerable regions through the formation of a local power structure certified by the center and with lines of authority again established.

The nationwide completion of provincial-level revolutionary committees represented not just the conclusion of a phase of the Cultural Revolution; it also set the stage for beginning a new phase of development, that of the rebuilding of the party. This was signaled by the major joint editorial on National Day, 1 October 1968, which was mainly devoted to the "new period" that the country had entered after the establishment of revolutionary committees.[4] Party rebuilding was put on the agenda as occupying "a very important position" in the ongoing transformation of the superstructure that was the object of the Cultural Revolution. This process soon thereafter was taken a very important step further when a new Central Committee Plenum, the first since the August 1966 Plenum formally endorsing the Cultural Revolution, was held from 13 to 31 October. Like the stacked 11th Plenum two years earlier, the 12th was an "enlarged" plenum, comprising Central Committee members, all members of the special Cultural Revolution Group, and "principal" officials of provincial-level revolutionary committees and of the PLA. In the first direct naming of the disgraced Liu Shao-ch'i after eighteen months of circumlocutionary attacks on "China's Khrushchev," the plenum resolution declared that Liu's "bourgeois headquarters" had been "finally smashed" and that the Cultural Revolution had won a "great and decisive victory" with the completion of the formation of provincial revolutionary committees. Most significantly, the plenum decided that "ample ideological, political and organizational conditions have been prepared" for convening the 9th Party Congress, which was to be held at "an appropriate time."[5]

The plenum communiqué used a standard formula of the Cultural Revolution in calling Mao Thought the Marxism-Leninism of "the era in which imperialism is heading for total collapse and socialism is advancing to worldwide victory." It did not, however, call Mao the greatest Marxist-Leninist of the era or one who had raised Marxism-Leninism to a new stage. This failure to renew the Cultural Revolution claims of Mao's ideological genius and universal authority accorded with an earlier important signal to the international Communist movement that was in harmony with the new stage in Chinese policy. On 18 September—that is, a day after the message signed by Mao to the Albanians proclaiming "a historic new stage" of struggle—Peking with great fanfare released for the first time an inscription written by Mao six years earlier for a visiting group of Japanese workers.[6] On the same date in 1962, NCNA had reported that Mao had received the Japanese, but the inscription was not published nor the anniversary commemorated until now, six years later. Moreover, the fact that this was not a quinquennial anniversary made it all the clearer that Peking's purpose was to deliver an important message to the world rather than to commemorate a hitherto obscure event. The inscription itself was deceptively simple: "The Japanese revolution will undoubtedly be victorious, provided the universal truth of Marxism-Leninism is really integrated with the concrete practice of the Japanese revolution." The significance of the inscription's release at this time was that it signaled a significant withdrawal from the Cultural Revolution's claim of the universal validity and efficacy of Maoism for all countries. Note that the inscription referred to the universal truth of Marxism-Leninism, with no reference to Mao Thought. Moreover, it stressed the need for the concrete application of that universal truth to the practical conditions of a given country. This represented a major retreat from the universalist formulation used by Lin Piao at the height of the Cultural Revolution in defining the global reach of the Maoist writ.

Here was another instructive case of the contextual significance of an act, in this case the act of belatedly publishing an inscription that had been unexceptional in its time; the significance lay in the context, not the content per se. Amplifying the message, a *People's Daily* editorial on the same day[7] hailed the inscription as having "extremely great and far-reaching significance" not only for the revolutionary cause in Japan but also for "the revolutionary cause of the people of all countries throughout the world." Emphasizing the fundamental importance of close integration of Marxist-Leninist theory with the concrete situation of any given country, the editorial made it clear that this principle was applicable to China and Japan "and the revolution in all other countries as well." There was no equating of Mao Thought with Marxism-Leninism or any claim that dissemination of Maoism throughout the world was advancing world revolution; rather, Mao's particular merit consisted in his application of Marxism-Leninism to the concrete conditions of China itself. By implication, this in large measure retracted the universalist message of Maoist people's war Lin had espoused in his celebrated tract of September 1965.

The release of the inscription confirmed earlier indications that the Chinese were pulling back from the Maoist evangelism of recent years. On 12 April 1968, the *Peking Review* carried the last of its regular feature columns headed "Mao Tse-tung's Thought Lights the Whole World," which had spread the Maoist tidings by reporting foreign tribute to Mao Thought and recounting efforts across the globe to propagate Maoism.[8] There were also reports circulating in the Communist movement, and publicized by the Soviets, that the Chinese were notifying their followers that this sort of evangelism was no longer desirable. But the great importance of the release of the inscription and the accompanying editorial was that Peking was now prepared to go on record with an authoritative signal to the international Communist movement on a sensitive matter of Cultural Revolution politics. That Peking could hold a plenum to put a party congress on the agenda confirmed still more authoritatively that conditions were propitious for the new initiatives being taken during that period. The New Year's Day joint editorial ushering in 1969[9] also confirmed the sense that a new stage of development had been reached, noting that the Cultural Revolution had won a "great, decisive victory" and that the new year would see the 9th Party Congress. Significantly, the editorial stressed themes of moderation and unity, notably in the name of a recent directive from Mao decreeing that the target of political struggle must be narrowed. It also inveighed against factionalism, the characteristic product of Cultural

Revolution struggle. Unlike the 1968 New Year's Day editorial, this one did not characterize Mao Thought as Marxism-Leninism of the present era, though it did proudly conjoin Mao Thought and Marxism-Leninism in recalling that the 1960s had been a period of great struggle against modern revisionism. Peking may have been cutting its losses from Cultural Revolution excesses, but it was not conceding anything to the Soviets in the international Communist movement.

It will be recalled that, in the second half of September 1968, there had been a flurry of authoritative Chinese comment dealing with foreign affairs. *People's Daily* and *Liberation Army Daily* had carried editorials on the significance of the release of Mao's inscription for the Japanese workers, and likewise on the meaning of Albania's withdrawal from the Warsaw Pact; and a joint editorial had been devoted to hailing the arrival of the Albanian Defense Minister, whose visit was punctuated by Chinese charges of border tension. All of this represented a sharp upsurge in editorial comment dealing with international topics; previously that year, there had been only a handful of editorials discussing foreign affairs, and those dealt mainly with Vietnam. Significantly, what now happened was that the rise in attention to international affairs was accompanied by a notable reduction in the obsessive concern with the Vietnam War that had prevailed in recent years. This was not as paradoxical as it may seem on its face, for the rising hostilities in Vietnam had set the conditions for the Maoist leadership's adoption of an essentially isolationist posture in foreign affairs. If now Peking was taking new initiatives and defining a new stage in international relations, it would appear that a new assessment of the Vietnam situation was also in order, with consequences as profound as those of the conflicting assessments of the meaning of Vietnam in 1965. And it is worth a reminder here that among those earlier consequences had been important effects on the Chinese leadership situation.

On 19 October, while the Party Plenum was in session, Peking for the first time broke its public silence on the five-month-old Paris talks on Vietnam. An NCNA report rounding up Western press speculation about a new United States initiative—centering on a total halt to the bombing of North Vietnam—noted that the Paris talks had begun on 13 May and that 26 meetings had been held thus far.[10] Since the opening of the talks, the Chinese had insistently disparaged "peace talk schemes," but, until now, they had never openly mentioned the talks or acknowledged Hanoi's agreement to enter into negotiations; in fact, the Chinese indicated their displeasure by deleting Hanoi's own references to the talks when reporting North Vietnamese comment. Peking had denounced President Johnson's 31 March announcement of a partial bombing halt (and of his decision not to run for re-election) as a "plot of inducing peace talks," and subsequent Chinese comment had warned Hanoi that the United States' desire for negotiations was "a big fraud." But Peking's new approach in the autumn of 1968 was given even more startling expression when it carried the text of Johnson's 31 October announcement of a total bombing halt as well as the DRV's 2 November statement agreeing to an expansion of the Paris talks to include representatives of the Saigon regime and of the Viet Cong. This highly significant move by Peking, reversing the adamant hostility to Vietnam negotiations hitherto prevailing, signaled a new flexibility and an important new sense of openness to the possibility of a negotiated settlement. Even more important, it implied a new assessment of U.S. intentions—or perhaps more properly, of U.S. capabilities—in Indochina in particular and in Asia in general. Further evidence on this score was soon forthcoming in Peking's reaction to the American presidential election on 5 November 1968.

Peking's account of the election, in an NCNA dispatch three days later, pointed out that Richard Nixon was elected president (and here the punctuation was revealing) "after he called for the necessity to 'reduce our commitments around the world in the areas where we are overextended' and to 'put more emphasis on the priority areas,' namely, Europe and other areas" (the interior quotations being Nixon's words). Of all the flood of campaign oratory, only this fragment was used by Peking to explain the election results. The Chinese were, in effect, inviting the victorious candidate to read the results as a mandate for reversing the process of engagement in Indochina and to concentrate American resources and interests in areas like Europe, where U.S. interests collided with those of the Soviet Union rather than of the Chinese. NCNA also pointed out that the swapping of horses while crossing a turbulent stream, meaning the switch in parties occupying the White House while the Vietnam War was raging, was "a striking manifestation of the deep crisis" enveloping the United States. Interestingly, there was no personal mockery of Nixon; rather, in indulging in the expected ideological disparagement of the bourgeois election process, NCNA directed its mockery at the political parties ("the U.S. monopoly capitalist groups" could not but put "another tool of theirs—the Republican Party—into the White House").[11]

During this period, the Chinese were falling silent on the Vietnam question, apparently waiting to see how the negotiating process would evolve and meanwhile abandoning the obstructionist approach that had so long characterized Peking's treatment of the question. The new mood was also reflected in Peking's reaction to the annual vote on the China representation question in the United Nations. An NCNA report on 21 November said the adverse vote on the PRC's claim to the China seat showed the "political and moral bankruptcy" of the United Nations and proved the validity of Peking's assessment of that body as "nothing but U.S. imperialism's tool for aggression." But more significant was what was omitted from Peking's previous reaction to the same result. The 1967 vote had prompted a more authoritative reaction, a *People's Daily* "Commentator" article, which proclaimed Peking's disinterest in joining the organization and threatened the creation of a "new and revolutionary United Nations" if the existing body was not reformed. Now the Chinese had quietly retreated from that more revolutionary posture and began preparing the ground for promoting a favorable vote in the future.[12]

As has been seen, Peking in its reaction to the U.S. presidential election result had in effect invited the incoming administration to rethink the nature and extent of American political and military commitments in Asia. Now, in the most striking sign thus far of the new flexibility in Peking's stance, a direct overture was made to the new administration to explore the possibility of putting the Sino-U.S. relationship itself on a new and more promising footing after the unrelieved rancor of recent years. This overture provides a most interesting case study in the signaling process; and though, as will be seen, the incipient movement toward a new relationship proved abortive in the short term, it illustrated the complexity and delicacy of the signaling process, with the contingencies involved in a meeting of minds of the parties themselves as well as their ability to sustain their respective mandates through the vicissitudes endemic to the process. If nothing else, this particular case affords an instructive lesson in the need for contextual analysis of the terms of reference in the message being communicated; the content of the message taken more or less in its own terms out of context proves inadequate at best and possibly maleficent in effect.

Peking's signal to the United States took the form of a statement, issued in the name of the PRC Foreign Ministry spokesman, disclosing that the Chinese envoy in Warsaw had, on the previous day, 25 November 1968, written to the American ambassador to propose that the next, 135th session of the Warsaw talks take place on 20 February.[13] The statement went to tortuous pains to shift the responsibility from the Chinese for postponing until February the meeting that had been scheduled for November, but in doing so it made clear that its real purpose was to convey Peking's readiness to resume the talks with representatives of the new administration, which, as the statement pointed out, would have been in office for a month by the date now being proposed. Peking also took the occasion to enunciate two principles which, according to the statement, the PRC had consistently followed in the thirteen years thus far of the Sino-American ambassadorial talks. The first principle embodied the demand that the United States withdraw its military forces and installations from Taiwan and the Straits; the other called on the United States to join in an "agreement on the five principles of peaceful coexistence." The statement also indicated that the Chinese were interested in exploring fundamental issues. It complained that the United States had been putting the cart before the horse by persistently "haggling over side issues"—an allusion to United States proposals to improve the atmosphere through various exchange programs—and it warned that a continuation of this approach would mean that no result whatsoever would ensue from the talks "no matter which Administration assumes office" in Washington.

There were two contextual factors important to an understanding of Peking's message, the first its timing, the other its invocation of peaceful coexistence as a goal. The Chinese had felt constrained to rebut U.S. accusations that they were avoiding the 135th meeting because they had already once before postponed it. NCNA, on 28 May 1968, had disclosed that the Chinese chargé d'affaires in Warsaw had ten days earlier written to the American ambassador proposing that the 135th meeting scheduled for 29 May be postponed until November "as there is nothing to discuss at present." It was quite understandable that the Chinese wished to avoid meeting in May, for that would have seemed to compromise their total opposition to the Paris talks between their North Vietnamese comrades and the Americans, which began on 13 May. By November, however, the context had significantly changed, Peking having not only acknowledged the legitimacy of the Paris talks but also having taken note of a major breakthrough. It was therefore appropriate, in such a context, for the Chinese themselves to seek to enter on a negotiating track with the United States to explore the prospects for a new relationship.

It might be argued that, after all, Peking was merely returning to talks that had been taking place even during the most virulent phases of Sino-U.S. hostility, having only avoided meeting with the United States for a time in order not to detract from the line of opposition to the Paris talks. Having now resigned themselves to these talks, the Chinese might have seen no harm, but no great prospects either, in resuming the Warsaw talks. Such a view of Peking's November 1968 overture not only overlooks the general context indicating a new flexibility in the Chinese position, but it also fails to give proper weight to Peking's proposal for an agreement on peaceful coexistence. Again, as in the case of the charge of Soviet troop concentrations, for another example, it was not a matter of introducing something wholly new that constituted the signal; the proposal on peaceful coexistence, coupled with the demand for American military withdrawal from Taiwan, had marked Peking's position in the Sino-American ambassadorial talks from their beginning in the 1950s. The significance now, however, was that Peking was *reviving* this proposal, which represented a striking shift in Peking's position since the mid-1960s of intransigent opposition to the notion of détente with the United States and to the possibility of a negotiated settlement of such issues as Vietnam. It should be noted, for example, that Foreign Minister Chen I's speech marking the Sino-Soviet treaty anniversary in February 1965 flatly ruled out peaceful coexistence with the United States as "out of the question."[14] To put the November 1968 overture further into perspective, it should also be noted that the last previous Chinese statement on the Warsaw talks—the unprecedented publication, on 7 September 1966, of the PRC ambassador's statement at the Warsaw talks that day—coupled the demand for U.S. withdrawal from Taiwan with a reaffirmation of Peking's intent to "liberate" the island. Moreover, there was no reference then to peaceful coexistence. The release of the ambassador's statement was accompanied by a press conference at which he made clear that the Chinese were reacting to Soviet propaganda baiting Peking for carrying on the Warsaw dialogue with the United States while posing as proponents of revolutionary armed struggle as the only means for dealing with the enemy.[15] Now, in late 1968, having acknowledged the legitimacy of the talks seeking a Vietnam settlement, Peking was signaling its intent to reopen the dialogue with a new administration to explore the implications of a post-Vietnam settlement.

It will be noted that Peking's moves since the invasion of Czechoslovakia followed parallel lines in the domestic and international spheres, suggesting that a leadership consensus had formed that permitted new departures in what amounted to a post-Cultural Revolution phase of policy. As has been seen, the initiatives taken in the autumn of 1968 represented in various important respects departures from policies characteristic of the Cultural Revolution; this in turn raised the possibility that the leaders with particular vested interests in those policies would look with the most skepticism and unease at the new departures. Consider, for example, the role that evangelical propagation of Mao Thought had played in raising Lin Piao to a position of eminence second only to the Chairman himself. The contraction in the presumed scope of Mao's ideological authority now being signaled to the world might also seem to Lin and his followers a warning signal of his own reduction in status. On the other hand, the new directions being taken during this period were closely associated with Chou En-lai, either as the spokesman or as the patron of the interests most directly involved, such as the Foreign Ministry, the institution likely to acquire a more active role as the PRC adopted a more flexible approach in international affairs. In short, a move from the isolationism prefigured in Lin's 1965 tract on people's war toward more active involvement in foreign affairs had an important impact on the positions of crucial figures in the Peking leadership.

The October Party Plenum, the importance of which was indicated by its decision to put a party congress on the agenda, provided an occasion for extensive deliberations by the Chinese power elite. The Plenum's significance was further underscored by the republication at this time of Mao's report to a Central Committee Plenum in March 1949. An authoritative joint editorial in *People's Daily, Red Flag,* and *Liberation Army Daily* on 25 November 1968 quoted a recent Mao instruction, "Historical experience merits attention," in explaining that study of the 1949 report was of great significance for carrying out the tasks of the recent plenum. The editorial suggested that there was a parallel between the two situations—the March 1949 Plenum a few months before the establishment of the new regime, and the October 1969 Plenum heralding the convocation of a new congress (and, in fact, there was a comparable length of time before the consummations of the respective plenums). Though the editorial did not address itself to this aspect, one message being conveyed in the republication of the 1949 report may have related to the overture made to the United States that day. Republication of a work from the Maoist canon indicates that it is meant to be reread for current edification, and the reader would find in it a passage explaining that the Chinese Communists were preparing for new negotiations with the Kuomintang. In this connection Mao's report said: "We should not refuse to enter into negotiations because we are afraid of troubles and want to avoid complications, nor should we enter into negotiations with our minds in a haze. We should be firm in principle; we should also have all the flexibility permissible and necessary for carrying out our principles."[16] This may have been the sort of mandate the Chinese leadership had now handed to Chou and his associates for exploratory negotiations with the United States.

It was against this background that a new administration entered office in Washington, and a measure of the importance Peking attached to the event can be found in the unprecedented scope and authoritativeness of its commentary devoted to the Nixon inauguration. The thrust of the commentary was to picture the United States as in a state of ineluctable decline into a sea of troubles at home and abroad. Displaying a confidently mocking style that contrasted with the strident paranoia marking Peking's diatribes during the Cultural Revolution, the Chinese portrayed the incoming administration as trapped in a morass of crises and antagonisms born of bankrupt domestic and foreign policies. However, behind all the propaganda sound and fury that seemed to protest too much, Peking conspicuously avoided foreclosing judgment on the crucial issues bedeviling Sino-American relations that would be expected to figure in the impending Warsaw talks. Moreover, Peking published the text of the Nixon inaugural address, which contained an expression of hope that no people in the world would live in "angry isolation." By avoiding definitive judgment on the substantive issues, Peking was leaving open the possibility that steps might be forthcoming for alleviating the angry isolation that had divided the United States from mainland China.

In the most authoritative commentary on the inauguration, *People's Daily* and *Red Flag* carried a joint "Commentator" article on 27 January entitled "Confession in an Impasse—A Comment on Nixon's Inaugural Address and the Despicable Applause by the Soviet Revisionist Renegade Clique."[17] The article called the inaugural address "excellent teaching material by negative example," which showed clearly "the very weak, paper-tiger nature of U.S. imperialism." In the most direct reference to Sino-American issues, "Commentator" responded to the new President's professed aspiration to be a peacemaker by challenging the United States to withdraw its troops from the Taiwan Straits, Vietnam, and all other places they had "occupied." (The call for withdrawal from the Straits was anomalous; the standard formula was for withdrawal from Taiwan *and* the Straits. This may have been an artful way of inviting the United States to show its *bona fides* by ending the Seventh Fleet's patrolling of the Straits, a practice which was in fact abandoned early during the Nixon Administration.) This challenge to a declining power to withdraw from its forward positions abroad was consistent with Peking's suggestive use (cited above) of candidate Nixon's pledge to reduce the United States' overextended commitments. Similarly, after the inauguration, Peking quoted the newly appointed National Security Adviser, Henry Kissinger, as acknowledging that U.S. global strategy had been a failure and that the United States was "no longer in a position to operate programs globally."

Peking's willingness to skirt the difficult issues was made the more conspicuous by its reaction to President Nixon's first press conference, on 27 January, when he said his administration would continue the opposition to Communist China's being seated in the United Nations. As for the scheduled Warsaw meeting, he declared that there would be no change in U.S. policy until Peking demonstrated a change on its part regarding major substantive issues. If, in fact, there had been no change, or at least incipient change, in

Peking's approach, these remarks by the new President would have been seized upon and flaunted by Peking as showing that Washington was obdurately hostile to the PRC. Significantly, however, Peking's account[18] of that press conference made no direct mention of his remarks on the United Nations or the Warsaw talks. In sharp and revealing contrast was Peking's treatment of similar remarks in a policy statement by Japanese Premier Sato on the same day as the President's press conference. As the Chinese put it,[19] Sato "publicized his dogged intention to persist in the policy of hostility toward China" and "made the outrageous demand that China 'change' its attitude toward the reactionary Japanese Government." President Nixon had likewise made a change in U.S. policy conditional on "changes of attitude" on Peking's part, but Peking had not elected to fling his words back in his face as it did to the Japanese leader.

There was one aspect of the new administration's policy orientation that Peking viewed with particular concern and apprehension, namely, its posture toward the Soviet Union. A theme running throughout Peking's commentary on the inauguration was that, under the new administration, Washington would continue to give a high priority to an improvement of relations with Moscow, or what the Chinese were wont to call U.S.-Soviet collaboration in the interests of global domination. Typically, the arms control area provided the focus for Peking's apprehensions over Soviet-U.S. relations, and the Chinese made a point of noting that, on the very day of the presidential inauguration, the Soviets issued a policy statement on arms control that included a reassertion of interest in strategic arms limitation talks (SALT). The Chinese also expressed sharp resentment over Soviet moves to establish a good working relationship with the incoming administration. A *People's Daily* article two days after the inauguration, entitled "Shameless Claque,"[20] described how, in an "utterly nauseating" way, the Soviets had hastened to "fawn on" the new President. In a significant formula that departed from the usual characterization of Moscow as the "number one accomplice" of Washington, the article referred to the "two number one enemies, U.S. imperialism and Soviet revisionism." In later comment, particularly in response to the President's first two press conferences, Peking took special note of the new administration's professed desire to establish a linkage of issues involving the U.S.-Soviet relationship. The Chinese pointed out in particular that Washington hoped to use Moscow's desire for SALT as the key link in a package deal covering several issues.

How Peking perceived the Soviet-U.S. relationship—or more significant methodologically, how Peking portrayed that relationship—has been a central element in its definition of the possibilities for taking a relatively flexible approach toward one or the other of the superpowers. Previously, a hard line toward the Soviets had been correlated with a portrayal of them as the main accomplice of the United States and thus as standing condemned for having traitorously put their relationship with the United States above their obligations to the Communist movement. Now, the Soviets having been further anathematized as "social-imperialists" and defined as a number-one enemy in their own right, the question was shifting to one of whether there was a sufficient divergence of interests between the Soviet Union and the United States to justify a flexible approach toward the latter. For the degree to which there was a perceived divergence rather than congruence of Soviet and U.S. interests, to that degree the proponents of flexibility toward the United States could make their case. As it developed, the movement toward improved relations with the United States was accompanied by precisely just such a shift toward the portrayal of divergent as compared with congruent elements in the Soviet-U.S. relationship. This provided an important indicator of Chinese policy changes.

In the midst of the stream of commentary on the presidential inauguration, Mao reappeared after being absent from public view since the time of the October Plenum. A big rally on 25 January 1969 was attended by Mao and a wide turnout of the Chinese leadership, which indicated that the elite were together in Peking and available for new deliberations in case of strains in the consensus reached at the plenum.[21] A very turbulent period was soon to ensue, but it is not possible to identify when the decisions contributing to that turbulence were made. A sign of trouble in Sino-U.S. relations appeared in Peking's reaction to a State Department announcement on 4 February that a Chinese diplomat in the Netherlands, Liao Ho-shu, had defected and been given asylum in the United States. A PRC Foreign Ministry spokesman's statement two days later[22] called this "a grave anti-China incident deliberately engineered by the U.S. Government." In a hint that a cloud hung over the Warsaw meeting scheduled for 20 February, the statement mentioned that a strong protest had been lodged with the American ambassador in Warsaw. Moreover, the Chinese used the occasion to take a personal slap at the President, saying that the incident showed that in being "hostile" to the PRC he and his predecessor were "jackals of the same lair without the least difference." President Johnson's overtures for relaxing tensions with Peking had been resoundingly rejected, and the coupling of the old and new administrations on this score did not bode well for the renewal of the Warsaw dialogue. Nor did a *People's Daily* article on 18 February,[23] belatedly reacting to President Nixon's 27 January press conference (and making up for having passed up the chance in earlier reaction to express hostility), which sneered that "Nixon, Johnson and all other heads of the U.S. imperialist governments past and present are agents of the U.S. monopoly capitalist class and blood-suckers of the working people." Finally, in a Foreign Ministry spokesman's statement released by NCNA at midnight on 18 February, Peking ended the suspense by announcing that it was "obviously most unsuitable" to hold the Warsaw meeting scheduled for the 20th in view of "the current anti-China atmosphere" created by the U.S. Government over the defector. Sharpening still further the portrayal of continuity on China policy between the new administration and its predecessors, the statement claimed that Washington's behavior in the defector incident revealed the "vicious features" of the Nixon Administration, which had "inherited the mantle" of its predecessor in "flagrantly making itself the enemy" of the Chinese people.[24]

Thus had the signals pointing to new movement in Sino-American relations been drowned out by a reversal of events, whether that reversal resulted from a lack of sufficient feedback signals from a U.S. administration that inadequately comprehended the Chinese signals, or from a collapse of a recent and fragile consensus in the Peking leadership, or (more likely) from a combination of these reasons. Given the two decades of Sino-American antagonism, exacerbated by the bitter Vietnam conflict, it is not surprising that the signaling process between Peking and Washington, with a new administration entering office, would require some false starts and tentative probes before any mutuality or even calculated parallelism could be perceived and communicated. As it happened, in the course of its first year in office, the Nixon Administration undertook a carefully calibrated series of steps to signal to Peking that such a process had real and promising prospects in a post-Vietnam environment. As has been seen, the months after Czechoslovakia had been notable for the emergence of a political atmosphere in Peking that permitted the inception of new policy departures. But, compared with the United States, where a war-weary populace was receptive to new foreign policy directions to be undertaken by a conservative leadership, the Chinese domestic situation remained fraught with instability and conflict. The October CCP Plenum, in setting the agenda for the "new period" following completion of the establishment of revolutionary committees in all provinces, had also set the stage for another round in the recurrent power struggles that were to become a legacy of the Cultural Revolution. In their most basic sense, these struggles revolved around the question of the extent to which the Cultural Revolution forces were to be curbed and displaced by newly revived forces that themselves had been removed or curbed during the Cultural Revolution.

In the wake of the October Plenum, certain themes began to appear in polemical discourse reflecting opposition by Cultural Revolution forces to the new policy lines evolving that autumn. In particular, exploiting the fact that the Plenum had formally condemned Liu Shao-ch'i by name, these polemical themes used Liu and his actions in the past as a surrogate for the new policies. Thus, it was argued that Liu had opposed Mao's line on revolutionary armed struggle and had favored political means, and that he had been soft on the United States rather than promoting Maoist liberation struggles. Apart from their content, with its provocative resonance for current developments, these polemical sallies were also notable for their provenance. One that appeared in *People's Daily*, for example, was attributed to a member of the Army's Higher Military Academy, an indication that Lin Piao's own vested interests were involved. A series of similar attacks was carried in the Shanghai paper *Wen Hui Pao*, which three years earlier had been the source of the opening shot of the Cultural Revolution by Yao Wen-yuan, one of the Shanghai leaders who were the ideological spearheads of the Cultural Revolution.[25]

When Mao returned to view in January after being absent since the plenum, he undoubtedly was lobbied intensively by the Cultural Revolution forces, alarmed over the implications of the new policy directions for their own interests. In addition to the cancellation of the Warsaw meeting, an obvious setback for the move to explore a new relationship with the United States, the Chinese portrayal of the Nixon Administration became increasingly negative, an indication that the area of diplomatic flexibility available to Chou En-lai and his confederates was being sharply constricted. In particular, as regards the key indicator of the perceived Soviet-U.S. relationship, the Chinese portrayal now emphasized the elements of "collusion" while minimizing if not ignoring those of "contention." Contrary to the remarkable signal of a changed perception offered by Chou the previous September,[26] the portrayal of fundamentally congruent Soviet-U.S. interests and policies served—as it had during Lin Piao's ascendancy—the dual confrontationist line closely associated with Lin and the Cultural Revolution. What now came to pass, the gravely menacing Sino-Soviet border confrontation, put that line to a severe test, one that became entwined with the ongoing power struggle in the Chinese leadership. In the dark new mood emerging in Peking, the border clashes erupting in March 1969 were interpreted in the framework of a de facto Soviet-U.S. alliance directed against China, thereby underscoring that aspect of Peking's early commentary on the Nixon

Administration stressing the new President's interest in dealing with the Soviets. This judgment was reinforced by Nixon's 14 March announcement of the decision to proceed with plans for a limited antiballistic missile system, a decision that NCNA on 16 March—a day after a major Sino-Soviet border clash—called "further evidence of joint military opposition to China" by the United States and the Soviet Union.[27] Peking dwelt on what it stressed was the anti-China orientation of the planned ABM system, quoting Nixon as saying the United States and the Soviet Union had a common aim in deploying this kind of system to deal with China, and as calling for Soviet-U.S. talks on controlling the arms race in order to work together in coping with the potential Chinese threat. A few days later, NCNA linked the border clashes that month with the Kremlin's "anxious desire to enter into fresh and wider counterrevolutionary collusion with the new U.S. imperialist chieftain Richard Nixon."[28] Significantly, NCNA observed that, in almost every one of his press conferences since taking office, Nixon had offered some tidbits to the Soviets in his desire for closer U.S.-Soviet collaboration; even some Western papers were "amazed by the fact that the United States and the Soviet Union have entered into such passionate flirtation only two months after Nixon took office."

The developments since Peking's November 1968 overture to the incoming Nixon Administration illustrate well the complexities and contingencies of political communication and change. The language of Richard Nixon, who a year before his election had published an article bespeaking an openness to a new approach to the China problem,[29] now resonated with the anti-Chinese rhetoric of his past; this tough tone derived from domestic political considerations as well as from the compelling concerns of the U.S.-Soviet strategic relationship. As for Peking, those elements in the Chinese leadership looking to new departures in Sino-U.S. relations—and who had accounted for the November overture and the notably restrained reaction to Nixon's earliest presidential pronouncements—had not only failed to receive the reinforcing signals they needed, but they saw their tenuous mandate slip away under pressure from internal rivals anxious to preserve their own vested interests and shore up their challenged authority. The border crisis served initially to reinforce the dual confrontationist policy of the Cultural Revolution forces, and the CCP Congress in April 1969 raised that policy's foremost proponent, Lin Piao, to an unprecedented status in Communist history. These developments, however, contained the seeds for still further and more dramatic change, again involving complex interacting factors, which would lead to a new rethinking and reshaping of events that ushered in a new decade of international relations.

SIX

The Border Clashes

The lethal fire-fights along the Ussuri river in March 1969 that startled the world served more to punctuate than to initiate the process of grave deterioration in Sino-Soviet relations at the end of the decade. This is not to minimize the devastating effect of those eruptions; Moscow gave a count of 31 fatalities on its side in the first clash, on 2 March, and the clash thirteen days later was on an even greater scale. One can begin to appreciate the gravity of these events by imagining the effect of a shoot-out between Soviet and American troops in a Berlin crisis in which, say, 31 Americans were killed. Yet with all the tensions and crises in the Soviet-U.S. Cold War over Berlin and elsewhere, there never have been any such fire-fights, and rules of the game of Cold War were early evolved to ensure that local incidents and initiatives would not trigger an eruption whose consequences would be difficult to predict or control. The 4,500-mile-long Sino-Soviet border had been a troubled area throughout the 1960s, with tensions there matching the poisoned political relations; but there, too, a pattern of practices and unwritten rules had kept in check the suspicions and passions aroused by the deteriorating political relationship. What happened in the wake of Czechoslovakia was that those rules of the game could no longer contain the effects along the border of new political decisions whose genesis has been discussed above. For half a year in the middle of 1969, until important new political decisions permitted the observance of newly stabilizing rules along the border, the Sino-Soviet confrontation reached a peak of tension and instability that was fraught with the most menacing potential for spiraling out of control.

An attempt to assess blame for the outbreak of fighting is unrewarding, not only because of the contradictory claims of the two sides and the inaccessibility of the facts, but also, and more important, because it would presuppose an overall judgment as to the relative merits of the two sides' general positions on the border question per se. Thus, if one granted Moscow's claim to the contested islands along the riverine border and hence its right to push Chinese trespassers from Soviet soil (or waters or ice), then the blame for the shoot-outs would fall on Peking for having aggressively challenged Soviet jurisdiction over the islands and for resorting to force to make good that challenge. If, on the other hand, one accepted Peking's entitlement to the islands, then the Chinese had a right to react with force to the other side's forcible measures to eject them from areas rightfully theirs. A Chinese documentary film released a month after the Chenpao/Damanskiy[1] Island clashes in March 1969, which gave graphic evidence of Soviet practices to which even Moscow's own accounts testified, showed how Soviet border troops harassed and repulsed Chinese who entered into areas claimed by Moscow. These were longstanding practices—such as using high-pressure hoses to drive back the Chinese, ramming Chinese fishing boats with gunboats, pushing Chinese off the contested islands; whether they were justified or not depended on the merits of the territorial claims but, for several years, and especially from 1967 on, these practices had, in effect, set the rules of the game. It was when the Chinese decided—and this was essentially a political decision—to contest Soviet jurisdiction more aggressively, and ultimately with firepower, that the menacing instability signaled by the March clashes overtook the rules obtaining theretofore.

It is thus important to take into account the factor of timing, rather than to seek to assign responsibility, in understanding the eruption of border fighting. As has been seen, the invasion of

Czechoslovakia was a major turning point in international Communist affairs not so much because of the direct impact of the Soviet intervention as because of Peking's reaction to certain other parties' acceptance of the action as a measure to preserve the integrity of the socialist camp. Determined to give substance to their charge that the Soviet Union was a "social-imperialist" country and that, thus, there could no longer be a socialist camp, the Chinese elected to make a political issue of border tensions and of alleged Soviet border violations. This in itself must have served as a warning signal to the Soviets, though they were not at all eager to join issue politically with the Chinese. These considerations were reflected in an article in *Izvestiya* on 3 December 1968 by the Commander of the Soviet Far East Military District, who recounted recent military exercises in this border district in order to deliver a message of high vigilance and combat readiness.[2] He avoided referring to the Chinese directly, mentioning U.S. actions in Vietnam and Korea as the source of international tensions, but he noted that, during the time of the invasion of Czechoslovakia, the troops in the Far East Military District particularly showed their discipline and vigilance. The Chinese must have seen the relevance of that observation to their charges of Soviet overflights at the time of Czechoslovakia, and in fact the *Izvestiya* article prompted a polemical reply. This reply, by a PLA member in *People's Daily* on 23 January, took the occasion to renew the charges of Soviet troop concentrations along the border and intensified provocations against China as demonstrating the social-imperialist nature of the Soviets. Interestingly, the article was broadcast only in Radio Peking's Vietnamese service, another indication that the Vietnamese were a primary audience for Peking's message on Soviet social-imperialism.[3]

Against the background of the political developments after Czechoslovakia, it seems likely that Peking decided to translate the border issue into more aggressive patrolling and challenging of Soviet jurisdiction in disputed areas. This is certainly not to suggest, however, that Peking was reacting out of apprehension over Soviet intentions toward China, as distinguished from Soviet intentions toward the Balkans. To raise the border question and to challenge the Soviets more aggressively would hardly be the way to allay Soviet suspicions; rather, the Chinese would have wanted to minimize provocative acts if they were fearful of Soviet intentions in this region. On the other hand, the effort to deter the Soviets from further ventures in East Europe would be served by keeping them off balance with reminders of a troubled border on the Asian front, and aggressive border patrolling would give teeth to those reminders. Moreover, a more active challenge along the border would put teeth in Peking's charge of social-imperialism; if Peking's capability for action in East Europe was, perforce, limited, the long Sino-Soviet border afforded a directly accessible arena for carrying the challenge to the Soviets. And as has been seen, most notably in the shrill exasperation expressed by Chou En-lai in his highly significant address on 2 September 1968, favorable reaction to Soviet intervention in Czechoslovakia by the North Vietnamese and North Koreans gave the Chinese cause to mount such a challenge.

There is one sense in which Chinese concern over Soviet intentions on the Asian front can be made an intelligible part of an explanation of the border crisis. It may have been that some elements in the Chinese leadership expressed apprehension over Soviet intentions, and made a different kind of use of the other Asian Communists' reaction to Czechoslovakia, in order to argue that Peking would be prudent in these circumstances to moderate its anti-Soviet challenge. The existence of that kind of counsel would help account for the exceptionally pointed language used by Chou in September 1968, particularly his reference to the destruction of the socialist camp (an issue that had figured in an earlier Chinese leadership struggle involving policy toward the Soviets). It would also be the kind of counsel that would animate Mao to undertake another one of his virulent anti-Soviet campaigns. As has been adumbrated, the pattern of Chinese initiatives and moves in the wake of Czechoslovakia involved issues and interests that can be seen to have variously affected key figures in the Peking leadership. It should be no surprise, therefore, that a Mao-Chou axis was central to the restructuring of alliance and adversary relationships that was to mark the international environment in the new decade. Nor should it be surprising that Lin Piao, the principal victim of that axis in the domestic power struggle, would be branded as a "capitulationist" to Soviet social-imperialism.

For the newly animated anti-Soviet challenge to take the form of aggressive patrolling of disputed border areas, the icebound seasons along the Ussuri and Amur Rivers provided the most opportune times. According to subsequent Chinese accounts, the Soviets committed 16 intrusions into the Chenpao area from 23 January 1967 to 2 March 1969, the date of the first fire-fight there. Eight, or half, of those alleged intrusions were said to have taken place in the first two months of 1969 alone, an indication that this particular icebound season had witnessed a sharp upsurge of tension, which, in turn, suggested a mutually reinforcing increase in patrolling by the two sides. Even more significant, the Chinese said that, on 16 February, the Soviets threatened to use force if the Chinese patrolled there again, and that the Soviet Far East border guards entered into number-one combat readiness.[4] The Soviets have not challenged the accuracy of this scenario, though of course they would not acknowledge that their patrols amounted to intrusions. Clearly, the situation by that time had become explosive, with each side aggressively patrolling an island they both claimed, and now with the threat of the use of force contributing an incendiary element. A collision course had been set. On what particular day the collision would occur, and which side would open fire first, were matters that conceivably were the product of chance in a real enough sense. But the important point is that matters had reached such a pass as a result of basic political decisions in a very political context.

Each side, of course, blamed the other for having opened fire first and for having violated the other's frontier. The first to announce that a clash had occurred on 2 March was Moscow, and it was noteworthy that that announcement, in the form of an official protest note, was disseminated in translation over the TASS international service before it was carried in Russian.[5] That a translated version of the protest note was released before the Russian version revealed something about Moscow's priorities and concerns in publicizing the protest. The Soviets seemed to have reacted with shock, charging that the Chinese had ambushed the Soviet border guards, and quite likely the Soviets were genuinely surprised to learn that the old rules of the border game were no longer in operation. The Chinese version, consisting of a protest note accompanied by a preface taking cognizance of Moscow's earlier released note, was not broadcast until almost four hours after the TASS transmission. That the Chinese may thus have been reacting to Moscow's publicized protest rather than to the incident itself raises the possibility that Peking might not have intended to make a public issue of the clash. Such a possibility can surely be ruled out. Peking media characteristically have a relatively slow reaction time, and, once having learned that Moscow had released its own note, it was natural that the Chinese would preface theirs with a reference to the Soviet note. Moreover, the Chinese, in sharp and revealing contrast to the Soviets, immediately followed up the incident with a massive propaganda campaign that indicated they had every intention of making full use of it for major political purposes. Indeed, that would seem to have been the underlying motive for their challenging the old rules of the game along the border.

The preface to the Chinese note cited a Foreign Ministry spokesman as arguing that, even according to the "unequal" Treaty of Peking, which gave to Russia the area east of the Ussuri, Chenpao was indisputable Chinese territory and had "always been under Chinese jurisdiction." The Chinese were thus indicating that they were not reopening the territorial settlement in claiming the island but were simply exercising jurisdiction over what belonged to them according to existing (albeit "illegal") treaties. This also indicated the logic behind Peking's new border tactics, which involved aggressive patrolling of symbolically significant areas like Chenpao in order to establish a claim to jurisdiction. The note said that the Soviets, shooting first, killed and wounded "many" Chinese border guards, who were compelled to fight back in self-defense after repeated warnings had produced no effect.[6] An NCNA report on 3 March pointed out that the previous day's clash was "by no means an isolated incident" and that there had been "incidents of bloodshed on many occasions" but, in citing the 16 alleged Soviet intrusions into the island since January 1967, it indicated only that Chinese border guards on these earlier occasions had suffered injuries rather than fatalities.[7] It seems clear that the 2 March fire-fight had shattered the rules of the game that had contained tensions in the past.

It was on 3 March that a massive anti-Soviet propaganda campaign was unleashed in China, featuring nationwide demonstrations that involved what an official tally put at 260 million people by the 7th; there were also demonstrations each day outside the Soviet embassy, thus reviving a practice from the turbulent days of 1967, the most convulsive year of the Cultural Revolution. Also on the 3rd, the senior Vice Foreign Minister, Chi Peng-fei, used the occasion of a Moroccan embassy reception to deliver Peking's message that the clash the day before was "another big exposure of the aggressive nature of Soviet revisionism as social-imperialism." But it was on 4 March that Peking issued an interpretation of the incident in a framework linking the shoot-out with the charges of Soviet border violations that had been pressed in the aftermath of Czechoslovakia. Entitled "Down with the New Tsars," a joint editorial in *People's Daily* and *Liberation Army Daily*[8] developed the theme that the border clash demonstrated the social-imperialist character of the heirs of the tsars now ensconced in the Kremlin. The editorial used the argument that, even according to the 1860 Treaty of Peking, the island belonged to China, and that Chinese border guards had a sacred right to patrol in Chinese lands. It also revived the charge that the Soviets had massed troops along the border, and it claimed that there had been repeated intrusions into Chinese territory and airspace (thus taking account of the new dimension of contested parcels of land, whereas the charges in the previous autumn had spoken only of air intrusions). Warning the Soviets not to violate Chinese territory, the editorial issued a standard line for the Chinese position:

"We will not attack unless we are attacked; if we are attacked, we will certainly counterattack." In other words, the Chinese would patrol areas they claimed and would answer force with force if challenged by Soviet border troops.

For several days after the initial border clash, the Soviets made no move to match the Chinese campaign. The 2 March protest note was buried on page 4 of *Pravda* two days later, and very little was done to elaborate the Soviet case on the border dispute. TASS, on 4 March, referred to what it called Peking's "reply note" to the Soviet protest, observing that the Chinese note advanced "impudent claims to Soviet territory" but failing to join issue on the terms of the territorial dispute. TASS also took note of the mass demonstrations in China, linking "this new fit of anti-Soviet hysteria" with the impending Chinese Party Congress and the Mao group's need for a new anti-Soviet campaign to unite its supporters on a platform of "adventurism and extreme chauvinism." It was left to the trade union paper *Trud*, rather than the more authoritative party, government, or military papers, to provide additional details on what must have been a traumatic event. On 5 March, *Trud* carried a report from the Soviet Far East that, significantly, portrayed a quiet situation on the border and sought to convey a mood of calm and business-as-usual in that part of the country. While massive anti-Soviet demonstrations were sweeping China, Moscow reported only that rallies were held along the Soviet side of the border, a way of promoting vigilance and morale among the people directly affected.

Thus, there was a significant asymmetry between the Chinese and Soviet responses in the immediate aftermath of the first clash. After being quick in getting their account of the event to the world at large, the Soviets were notably reticent about the incident and were clearly desirous of localizing its impact and avoiding engagement in a broader territorial dispute. After several days of restraint, however, Moscow was constrained to reevaluate its approach in view of the exceptional scope of the ongoing Chinese campaign; the Soviets may well have come to the reluctant conclusion that continuing restraint would be taken as a sign of weakness, and they thereupon proceeded to take vigorous steps to signal the adversary that they were resolute and determined to repulse the Chinese challenge. The Soviet political and propaganda counteroffensive, which was very extensive and well-orchestrated, was launched on 7 March with a press conference sponsored by the Foreign Ministry for Soviet and foreign correspondents. In addition to buttressing Moscow's line that the 2 March incident blew up after a routine move to expel Chinese trespassers (what the Soviets were calling "established border procedures") was met by a carefully plotted ambush, the Soviet spokesman at the press conference revealed the first detailed casualty figures—31 dead and 14 wounded Soviet border guards. Significantly, the Soviets also took the occasion to introduce the theme of Chinese atrocities against wounded border guards, with photographs as evidence. On the next day, reports of a nationwide wave of protest meetings began to splash across the Soviet press. The belated campaign was in full swing.[9]

Also on 8 March, another important sign of a change of approach by the Kremlin appeared when *Pravda* carried an authoritative editorial article interpreting the border clash as serving Peking's "far-reaching aims" of a "radical reorientation" of Chinese policies and of "ultimately turning the PRC into a force hostile to the socialist countries."[10] Moscow was thus beginning to perceive the geologic changes being signaled by Peking's new challenge. But though the editorial article made passing references to Peking's "absurd territorial claims," there was still a conspicuous reluctance to join issue on the territorial question. One purpose of the campaign, however, was to put the Chinese on notice that the Soviets were determined to hold firm on what they regarded as the accepted border line. Appropriately, the military paper *Red Star* was most pointed in delivering that message. Half the front page of the 8 March issue was covered with reports of protest meetings, introduced with a warning against "any illusions" doubting Soviet determination to keep the border "inviolable and immune." A bit of rocket-rattling was in evidence in one of the reports, which quoted a statement made at a meeting in the Far East Military District that "the rocket troops showed at the important exercises just completed that the formidable weapons entrusted to them by the motherland for defense of the Far East are in strong, reliable hands. Let any provocateurs always remember this." *Red Star* followed up the next day with an editorial delivering (to quote its title) a "stern warning to the adventurists" against testing the Soviet borders.[11]

The Soviet campaign, which included demonstrations against the Chinese embassy in Moscow, prompted protests from Peking and a renewal of Chinese demonstrations "on an even greater scale" than earlier, according to NCNA, which on 12 March gave a startling figure of over 400 million participants in demonstrations in the past few days. Peking's most important move during this period, however, was the dissemination of a document by the Foreign Ministry's information department containing a detailed brief for Peking's claim to Chenpao. Entitled "Chenpao Island Has Always Been Chinese Territory,"[12] the document reviewed the historical background, especially the 1858 and 1860 "unequal" treaties by which tsarist Russia acquired the areas north of the Amur and east of the Ussuri, in developing the argument that the island belonged to China even according to those treaties. The key to this argument was that an established principle of international law, the thalweg principle, holds that the center of the main channel should form the boundary line determining the ownership of islands in navigable border rivers. The Soviets have not disputed the fact that the center of the main channel of the Ussuri runs between Chenpao and the Soviet bank of the river, but they have rejected the thalweg principle as governing the determination of ownership in this case. The Chinese further pointed out that in the secret 1964 border talks the Soviets had acknowledged that Chenpao and two nearby islands—which had also figured in incidents in recent years—belonged to the PRC. The Soviets were not likely to have been reassured by the reminder of the 1964 talks, which were involved in Peking's determined efforts during the latter phase of Khrushchev's rule to make political uses of the border dispute. This reminder was the more barbed in that the Foreign Ministry document recalled the Chinese warning in 1964 that they would have to "reconsider" their position "as a whole" if the Soviets remained adamant in rejecting Chinese demands. The meaning of this threat, which marked the outermost reach of the Chinese negotiating position, must be understood in terms of the multiple layers of the Chinese demands. As spelled out in the March 1969 document, Peking has made clear in talks with the Soviets that the treaties on the Sino-Soviet border were "unequal" treaties, but that the Chinese, nonetheless, were willing to take them as the basis for determining the boundary. But the Soviets refused to acknowledge that the treaties were unequal and sought (as the Chinese put it) to force Peking to accept "a new unequal treaty" which would legalize occupation of Chinese territory seized by crossing the boundary even as "defined by the unequal treaties." Particularly since the Brezhnev-Kosygin leadership had come to power, the document charged, the Soviets had sent large reinforcements to the Sino-Soviet border, "occupied still more Chinese territories," and conducted provocations along the border.

Peking had thus not only surfaced the 1964 border talks, which the Soviets could only have remembered with distaste and mistrust, but also updated the terms of reference in such a way as to show what lay behind the current border troubles. The reference to Soviet troop concentrations confirmed the link between the emergence of that charge in the previous autumn and the clashes of border patrols on the frozen Ussuri in March. Moreover, the claim that the Soviets had occupied still more Chinese territories explained why the Chinese had made a point of contesting Soviet control of Chenpao and other islands. Aggressive patrolling of the contested islands—to which the Chinese laid claim on the basis even of existing treaties—was a means by which Peking gave teeth to its argument that islands on the Chinese side of the main channel of the border rivers belonged to China. This was also a means of strengthening Peking's hand in determining the status quo, if that were to be used as the basis for an accommodation pending a negotiated settlement of the border issues. In the Sino-Indian border conflict, the Chinese had placed great emphasis on the line of actual control as the basis for an interim boundary alignment in the absence of a negotiated treaty settlement, and they had resorted to military force to compel Indian acquiescence in such an arrangement. Now, however, in the Sino-Soviet dispute, the Chinese found themselves in a more complicated situation which required them to blur the argument resting on an appeal to the status quo. On the one hand, they charged that the Soviets had disturbed the status quo by crossing the boundary as defined by the existing treaties; on the other hand, the Chinese resorted to aggressive patrolling of disputed areas in order to lay claim to exercising jurisdiction there. As a result there was an ambiguity in the Chinese position as regards what constituted the status quo; it had a juridical element in appealing to the thalweg principle to define the boundary line (while charging the Soviets with crossing the line), and it had a physical element in armed patrolling to make a show of actual control of contested areas. To the degree that the Chinese could exercise control over such areas, to that degree could the ambiguity be resolved and a less qualified appeal to the status quo be made. Hence the stubbornness with which the Chinese sought to establish and maintain a presence on Chenpao.

Moscow gave scant immediate public notice to the Chinese document, TASS on 11 March brushing it off as having voiced "mad territorial claims" and as having even demanded a revision of the border. TASS made no reference to the 1964 talks or to the treaties. During this period it was only in broadcasts to China that the Soviets broached the treaties, as in a broadcast citing the 1860 Treaty and an 1861 map "appended" to it showing that Chenpao belonged to Russia. But, despite this reticence on the territorial question, which Moscow preferred to have lie dormant, the Soviet campaign of protest and warning that began on 7 March indicated that border tensions had not subsided after the initial paroxysm of fighting. That the Chenpao situation remained unsettled was confirmed by a Chinese protest note on 13 March

The Border Clashes

detailing a series of Soviet incursions by border guards and armored vehicles as well as helicopters on six days from the 4th through the 12th, including two occasions on which Soviet vehicles entered the arm of the frozen waterway between the island and the Chinese bank.[13] According to a subsequent Soviet account, there was a meeting of border guard representatives on the 12th, one of the days the Chinese note said the Soviets had entered that arm of the waterway, and a Chinese officer allegedly threatened the use of armed force against Soviet border guards. Clearly the initial clash was followed by continued patrolling by each side, and a new collision was inevitable in the absence of instructions from one or the other side to its border units to avoid contesting the other side's patrolling. This patrolling was integral to Peking's challenge and to its determination to raise the border question anew; as for the Soviets, the nature of the second clash and of the follow-up polemical and political campaign reflected their determination to make a show of force which they hoped would dissuade the Chinese from taking their previous restraint as a sign of weakness.

As in the first clash, the second one, on 15 March, was first publicized by Moscow, which issued a Soviet Government Statement[14] charging that a large Chinese detachment, supported by artillery and mortar fire from the bank, attacked Soviet border guards defending the island. Subsequent Soviet accounts said that the Chinese forces numbered up to an infantry regiment with support units and that the Soviet units fought for more than seven hours to contain the Chinese attacks which "followed wave after wave." The period of maneuvering and warning since 2 March had set the stage for an expansion of the scale of combat to one substantially greater than the first clash. Apart from the numbers of combatants indicated in the reports, a measure of the scale was provided in *Pravda's* report on the 17th that a "mighty rain of artillery" came down on the Chinese during the "fierce" battle. According to the Chinese, heavy artillery shelling from the Soviet bank reached as far as seven kilometers inside Chinese territory. The Soviets were obviously intent on showing their muscle, but the Chinese were also intent on holding their ground and denying the Soviets uncontested control of the island.

The Soviets were more prepared both militarily and politically for the second clash, the full import of the earlier incident having brought home to them the need to appear decisive and forceful. This time the Soviet protest note was published on the first page of *Pravda* and *Izvestiya* on the day after the clash, and a campaign of protest meetings and portrayals of Soviet heroism and Chinese perfidy was immediately launched. A *Pravda* editorial on the 17th proclaimed in its title that "The USSR Borders Are Sacred and Inviolable," but it was left for broadcasts to China to spell out the implications of the show of strength. The Chinese were warned that the Soviet Union possessed superior military power, including nuclear-tipped missiles and other advanced weaponry, and this full arsenal of missiles could strike with great accuracy and devastating force. The Chinese were reminded that, having nothing to counter this might, "what they are capable of doing is letting millions of people die undefended." Moscow's diplomatic sensitivity to airing this kind of tough talk too directly was reflected not only in its being confined to broadcasts to China but also in its being limited to the purportedly unofficial channel of Radio Peace and Progress (the "voice of Soviet public opinion"). Moreover, Moscow avoided reacting directly to Western press reports of Soviet nuclear threats to China, an exception being a broadcast to Britain deriding "the provocatory false rumor" of possible nuclear action. Moscow was thus trying to limit the diplomatic impact of the border confrontation while conveying a message to the Chinese that further collisions along the border were fraught with the possibility of overwhelming reprisals.

Moscow also had recourse to its broadcasts to China for joining issue on the territorial question, another aspect of the confrontation that complicated Soviet diplomatic life. This reticence drew taunts from the Chinese, who claimed that Moscow did not dare to take up the question of the unequal treaties and the international law of boundaries. An NCNA account of the Soviet press conference held on 7 March seized upon signs of Kremlin indecision on how to treat the territorial question. According to NCNA and confirmed to some extent by recorded reportage of the press conference, Lt. Gen. P. Ionov of the Soviet border guards was cut off by Foreign Ministry spokesman L. Zamyatin while answering foreign reporters' questions about the 2 March clash. NCNA pointed out that Ionov was cut short when he blurted out that he could show the treaty confirming Soviet ownership of the island.[15] Apart from references to the treaty in broadcasts to China, Moscow preferred for some time after the two clashes to evade the territorial issue.

Eventually, however, Moscow had its juridical arguments in order and sought to take the diplomatic initiative on a matter that had shown no signs of going away. An authoritative Soviet Government Statement, delivered to the Chinese on 29 March and given full textual dissemination worldwide by TASS the next day,[16] dismissed the question of unequal treaties as a "concoction from beginning to end" but proposed that the formal discussions begun in 1964 should be resumed "in the nearest future." The statement indicated, however, that the Soviets intended to maintain their adamant position that the existing border as defined by Moscow was historically determined, both in the sense that it took shape long ago and passed along natural boundaries but also in that it was given legal form in the 1858 and 1860 treaties. Further, the statement contended that in 1861 the two sides had approved a map showing a demarcation line passing directly along the Chinese bank of the Ussuri in the Chenpao area. In another appeal to history that would acquire growing significance in coming months, the statement stressed that the historical picture would not be complete without mention of the "immense efforts and sacrifices" made by the Soviets in regaining control of the Far Eastern territories after the Bolshevik Revolution, and the "crushing rebuff" administered to the Japanese in later developments. This was a way of saying that, whatever juridical issues might be raised, the Soviets had made too great an investment to permit these areas to become subject to irredentist claims by others. And, as for the contested islands, the statement observed that, in the armed clashes with the Japanese, the issue was not the islands alone but claims against the borders of the Soviet Union and its Mongolian ally. With this suggestive historical analogy, Moscow made clear that it recognized the islands dispute as integral to the whole territorial question and thus as a matter of high stakes.

The terms in which the statement defined the border discussions reflected Moscow's determination not to give the Chinese a foothold for opening a territorial question. Thus, the statement recalled that the Soviet side has proposed "consultations" (*konsultatsii*) in 1964 for the purpose of "clarifying" certain sections of the border, which was a way of saying that the borders, having been juridically and historically determined, were not subject to further negotiation but only to a more precise demarcation of some stretches. The statement recalled that these so-called consultations, which had begun in Peking, were to have been resumed in Moscow on 15 October 1964, but the Chinese failed to resume them. What the statement neglected to mention was that the date came one day after Khrushchev was ousted at a Central Committee Plenum. Now, nearly five years later, developments had again reached a point at which talks on the border question were being proposed. Interestingly, the international party conference that was being prepared during the last weeks of Khrushchev's rule, and which had also been shelved, was now also again on the agenda.

SEVEN

The Ninth CCP Congress

In contrast to the massive campaign following the first border clash, Peking gave relatively little polemical and propaganda attention to the 15 March incident. The new pattern of differentiated reaction by the two sides—with the Soviets being the ones this time to make a big fuss over the event—gave further confirmation to the interpretation that the first clash resulted from a Chinese challenge to the existing rules of the game along the border, but that the second one resulted mainly from a Soviet determination to demonstrate the dangers of escalation inherent in the Chinese use of arms in border challenges. On 14 March Peking had released a *Red Flag* editorial containing Mao's latest directives and anticipating the imminent party congress, which was now the focus of attention. *People's Daily* on the 15th printed the important *Red Flag* editorial,[1] and on the next day the party daily gave greater prominence to the new directives and the editorial than to the latest border clash. The message being conveyed was expressed in the editorial's statement that the "tide of anger" toward Moscow's "unbridled aggressive ambitions" was flowing into a "tremendous upsurge" in all fields "to greet the party's ninth congress with concrete actions." The border dispute as such was barely in evidence in a Chinese Vice Premier's speech at a Pakistan embassy reception on 23 March, in which he spoke only vaguely about preparing against aggression by "imperialism and modern revisionism."[2]

The border question had thus been submerged into the broader interests of the party congress, which opened on the first day of April. In effect formalizing the results of the Cultural Revolution, which had delayed the convening of a congress until several years after it was constitutionally required, the 9th CCP Congress adopted a new party constitution and certified a new power structure reflecting the impact of the recent years' convulsions. The new constitution posited "Marxism–Leninism–Mao Tse-tung Thought as the theoretical basis guiding [the party's] thinking. Mao Tse-tung Thought is Marxism-Leninism of the era in which imperialism is heading for total collapse and socialism is advancing to worldwide vistory." It went on to claim that Mao had "inherited, defended and developed Marxism-Leninism and [had] brought it to a higher and completely new stage." This important ideological formula rectified the 1956 constitution's deletion of Mao Thought from the basis of the party's authority. It also marked the reversal of the trend signaled in September 1968 to reduce the scope of the claim in behalf of Mao's ideological authority; that claim was now again equal to the scope it had acquired at the height of the Cultural Revolution in 1967.[3] Another notable feature of the new constitution, and one intimately related to the claim in behalf of Mao's ideological pre-eminence, was its designation of Lin Piao as Mao's successor. For a party constitution to name a successor to the party chief represented a highly irregular departure from Communist practice, but it comported with the uses to which Lin had put Mao Thought in establishing his own authority as the leading expositor of what had now again been constitutionally sanctified as one of the sources of the party's authority. There was also, at this time, an institutional underpinning for Lin's special status, and this was the assumption of significant political power by military leaders in succession to the shattered party elite. The flow of political power into professional military hands after the decimation of the party apparatus was registered in the strong military representation in the Central Committee and Politburo produced by the 9th Congress.

As might be expected, Lin was given the honor of delivering the party's policy statement (the "political report") to the Congress on the first day.[4] And as also might be expected, Lin elaborated on the nature of Mao's ideological authority, particularly on the significance of the Cultural Revolution as "a new and great contribution to the theory and practice of Marxism-Leninism." Lin pointedly noted as "especially important" that the new constitution "clearly reaffirmed that Marxism-Leninism-Mao Tse-tung Thought is the theoretical basis guiding the party's thinking"—an allusion to the effect of the Cultural Revolution and of Lin personally in restoring to Mao what the 8th Congress had omitted from his ideological authority. In explaining the background of the Cultural Revolution, Lin's report pointed in particular to Mao's 1957 work "On the Correct Handling of Contradictions Among the People," a work coming in the immediate wake of the destalinization crisis in the Communist movement and representing a significant effort by Mao to come to grips with the challenge of that crisis. According to Lin's report, that work "for the first time in the theory and practice of the international communist movement" pointed out explicitly that classes and class struggle still existed after the socialist transformation of the ownership of the means of production had largely been completed. Lin's account of the origins of the Cultural Revolution was cast largely in terms of a conflict between Mao and Liu Shao-ch'i, a Manichaean scenario familiar in Communist historiography, but it could also be read in terms of the evolving ideological rivalry between Moscow and Peking (Liu's sobriquet of "China's Khrushchev" suggested the link between the two levels of conflict). Lin's account also graphically explained why such a drastic method as the Cultural Revolution was necessary to cope with the problems posed by class struggle in a socialist society. Lin quoted Mao as saying that various methods had failed until the discovery of "a form, a method, to arouse the broad masses to expose our dark aspect openly, in an all-round way and from below"—namely, the Cultural Revolution with its emphasis on mass criticism and its decimation of the entrenched party apparatus. How entrenched that apparatus had been was revealingly indicated in Lin's statement that the criticism "initiated" by Mao of the historical drama "Hai Jui Dismissed From Office" (a reference to Cultural Revolution ideologue Yao Wen-yuan's November 1965 polemical attack that first appeared in the Shanghai press) was directed at "the den of the revisionist clique—that impenetrable and watertight 'independent kingdom' under Liu Shao-ch'i's control, namely, the old Peking Municipal Party Committee." During Liu's stewardship of the party apparatus, Lin charged, "the broad revolutionary masses could hardly hear Chairman Mao's voice directly."

The Cultural Revolution had greatly deepened the ideological cleavage between Peking and Moscow, but there was also the confrontation of two major nation-states to be taken into account in a fundamental policy statement such as the congress political report. This was done in Lin's report principally through a revision of the basic "contradictions" shaping the international system, and here the milestone "general line" issued by Peking on 14 June 1963 provides a point of reference. As codified in Lin's political report to the congress, the four basic contradictions were now said to be between the oppressed nations on the one hand and imperialism and social-imperialism on the other; between the proletariat and the bourgeoisie in the capitalist and the revisionist countries; between imperialism and social-imperialism as well as among imperialist countries; and between socialist countries on the one hand and imperialism and social-imperialism on the other. The essence of this revised version was to place the Soviet Union ("social-imperialism") on the enemy side, thus formally registering the fundamental change in the nature of the Sino-Soviet conflict by the end of the 1960s. It also reflected Peking's renunciation of the socialist camp, which had no place in the new codification. The fourth in this set of contradictions was a revision of the previous version's first one, which was between the socialist camp and imperialism. The contradiction now occupying first place, one representing the national liberation movement, reflected the primacy accorded the politics of insurgency in the period since Lin's September 1965 tract on people's war. This was also reflected in the pride of place given in the political report's discussion of foreign affairs to revolutionary armed struggles in various countries.

In addition to a restatement of fundamental policy in the domestic and foreign spheres, the political report also dealt with that most pressing of current concerns, the border question, which, Lin noted, had "caught the attention of the whole world" since the clashes on the Ussuri in March. Having thus acknowledged Peking's awareness of the international community as a primary audience, Lin's report went to some length to present the Chinese position as calling for a negotiated settlement, in this connection noting that as early as August and September 1960 Peking had taken the initiative of proposing negotiations. As for the 1964 talks for which the Soviets had claimed the initiative, Lin charged that their "great-power chauvinist and social-imperialist stand" had led to the disruption of the talks because they refused to recognize the existing treaties as unequal and insisted that the Chinese recognize as belonging to the Soviet Union all the Chinese territory "which they had occupied or attempted to occupy in violation of the treaties." The latter charge, of course, related to the dispute over which the Chenpao fighting had erupted.

With all of the professed interest in a negotiated settlement of the border dispute, Lin's report at the same time expressed the intransigent mood prevailing in Peking at that time, a mood that could be called Maoist in the most personal sense. Lin disclosed that Premier Kosygin on 21 March had asked to communicate with the Chinese leadership by telephone—one of a long series of attempts by Moscow to remove the burden of the border conflict—but that the Chinese on the next day replied with a memorandum indicating that, in view of the present state of relations, it was "unsuitable" to communicate by telephone and the Soviets would have to conduct their business through diplomatic channels. This typically Maoist sauciness may have been prompted in part by lingering resentment toward Soviet gloating over the apparent effectiveness of Kosygin's direct cable to Chou in August 1967 demanding the release of a detained Soviet ship. At any rate, the rebuke to the Kremlin, now publicly aired, served to underscore Peking's determination to press an issue that the Soviets found distasteful and unbalancing. Lin's report contained Peking's first public acknowledgement of the 29 March Soviet statement, which showed Moscow in Chinese eyes to be "still clinging to its obstinate aggressor stand, while expressing willingness to resume 'consultations'." The political report, which was delivered on 1 April, adopted on the 14th, and published on the 27th, deferred an answer to the Soviet proposal by saying Peking was "considering its reply to this." The Chinese were evidently deferring a decision on their next move in the border conflict until after the results of the lengthy congress deliberations had become crystallized, particularly with respect to the new power elite emerging from the paroxysms of the Cultural Revolution.

Having presented their proposal on the border issue three days before the opening of the CCP Congress, the Soviets sought to follow up with signals aimed at creating desirable options for the Chinese in their deliberations. On 11 April, Moscow delivered a Foreign Ministry note that, couched in nonpolemical language, recalled that the two sides in the autumn of 1964 had agreed to continue the consultations and that the recent Soviet statement had favored resuming them. The note reiterated the proposal and invited the Chinese to reopen consultations in Moscow on 15 April or at another time in "the near future" convenient to them. The note expressed a hope for a "speedy" reply.[5] Five days later, or immediately after the proposed date for resuming the talks, Moscow sought to show its good will by observing a polemical moratorium toward the Chinese for nearly a week. During that period, Soviet media were virtually free of polemical comment on China, and Soviet broadcasts to China sought to press home the proposal to resume border consultations, as typified by a broadcast talk by a former Soviet military adviser in China and now an official of the Soviet-Chinese Friendship Society who professed a desire for friendship and cited the Soviet statement and note in evidence.

Peking effectively signaled its refusal to reciprocate in kind when it announced on 18 April that a documentary film, "Anti-China Atrocities of the New Tsars," would begin showing throughout China on the next day to portray graphically the background and recent developments of the border dispute. According to the announcement,[6] the film showed a map handed to the Chinese at the 1964 discussions that included in Soviet territory "more than 600 out of more than 700 islands" situated on the Chinese side of the center of the main channel of the Ussuri and Amur Rivers. The film also showed how the Soviets had been dealing with Chinese efforts to make use of contested areas: Soviet gunboats ramming Chinese fishing boats, high-pressure hoses being turned on Chinese fishermen, Chinese laborers being pushed into the water from the riverine islands. The massive Chinese demonstrations of the previous month were also shown, and this effort to arouse mass emotion was further reflected in subsequent press reports by workers, peasants, and soldiers expressing anti-Soviet indignation after seeing the film. The uses the film was designed to serve could be seen in pledges by these representatives of the masses to counter the Soviets by "concrete actions," such as promoting production and growing more grain as well as being prepared to "bury the invaders in the sea of people's war." Thus, at a time the Soviets were seeking to contain the border dispute, Peking was intent on giving it a broad scope.

Moscow waited until 22 April to take public note of the campaign pegged to the documentary film, which TASS interpreted as initiating "another round of the anti-Soviet campaign" and *Pravda* denounced on 23 April as "the most impudent and cynical slander of the Soviet people." In a Lenin Day address on the 22nd, I.V. Kapitonov declared that Chinese provocations on the border had aroused "anger and indignation" throughout the world and had received a "worthy rebuff." Repeating Moscow's earlier appeals for restraint along the border to avoid complications, he called for "urgent and practical" measures to normalize the situation. And he further put the onus on Peking to move toward talks by saying the Soviets were "waiting for a reply" to their proposals.[7]

Soon thereafter still another proposal was communicated. On 26 April, though not announced until six days later, Moscow proposed that a meeting of the joint commission on border river navigation be convened in Khabarovsk in May. Navigation questions were now becoming timely with the melting of the ice on the border rivers, as *Izvestiya* noted, rather tendentiously, in observing on 29 April that shipping would soon resume along the Amur, "that age-old Russian river." The 2 May TASS announcement[8] of the offer to hold the meeting recalled that the two countries had concluded an agreement in 1951 on rules of navigation along the border rivers and that 14 annual sessions of the joint commission had been held. TASS noted that the most recent session had been in 1967, forbearing from mentioning that Moscow had publicly charged in August of that year that the meeting had been undermined by Chinese insistence on introducing the question of border delimitation, a question not within the commission's competence. TASS now pointed out, however, that the Soviet side had suggested in 1968 that the 15th session be held in Khabarovsk that year but the Chinese "avoided taking part." While the TASS announcement carefully avoided broaching the border dispute, its presence hovered over other references by Moscow to the new navigation season then approaching. Thus, a Radio Moscow report on 10 May of the departure of a vessel from Khabarovsk bound for Iman on the Ussuri noted that this would be the first craft of the season to pass through the waters along Damanskiy (Chenpao) Island, though the clashes there were not mentioned. On 13 May, a broadcast describing the launching of the season's first convoy bound for Iman made a point of citing two anniversaries: that it was 325 years since the first group of Russians discovered the Amur, and 115 years since the first paddle steamers flying the Russian flag began regular navigation along the Amur and the Ussuri. Age-old Russian rivers.

Though Moscow had previously made clear its opposition to raising the border question in the navigation commission, an agreement to reopen that forum, however narrow its scope, would be directly relevant to the border conflict as a sign that the two sides were willing to seek provisional measures to regulate the situation and hence, in effect, to defuse the territorial issue. Thus it must have come as a favorable omen to Moscow that the Chinese agreed to convene the joint commission, though characteristically they counter-proposed that this take place in June. It was also significant that the Chinese reply, on 11 May, took pains to correct the public record on the failure to hold a session in 1968, thus indicating Peking's awareness of the international audience and its concern not to let Moscow gain credit as the conciliatory side in the dispute. According to the Chinese version,[9] they were actively preparing to attend a session in 1968 as proposed by Moscow but at that time they were "slandered" by Foreign Minister Gromyko in a policy statement to the Supreme Soviet on 27 June 1968.[10] This amounted to a warning to the Kremlin not to try to press the Chinese into a corner by signaling to the world that it was Peking that was being intransigent and unreasonable. This warning was further underscored by the observation in the Chinese reply that Moscow's account of the 1968 episode "inevitably arouses suspicion of the Soviet side's sincerity" about holding a new session of the commission. In this complex signaling process, Moscow got in a warning of its own when responding on 23 May to the Chinese reply.[11] While setting 18 June as the date for opening the commission's 15th session, Moscow took exception to the Chinese version of the 1968 episode and, in this context, alluded to the question of border delimitation by observing that the 1967 session ended without result because of the Chinese attempt to put on the agenda questions beyond its competence. In other words, Moscow was intent on finding ways of defusing border tensions without having to reopen territorial issues. But that the navigation commission meeting would not be so tailor-made for Moscow's purpose was indicated in the Chinese reply on 6 June, accepting the date Moscow offered.[12] Now challenging the Soviet account of the 1967 session, the Chinese charged that the Soviet side had refused then to follow the established practice of entering into the minutes the two sides' divergent views and "thus broke up" that session. According to the Chinese, Moscow was distorting the 1967 events in an attempt "to place in advance obstacles to the coming 15th meeting"; if it was sincere in wanting to make the coming meeting a success, then "it must change its wrong attitude and earnestly discuss all the questions put forward by the two sides concerning navigation" on the border rivers. This formulation carefully left open the possibility that the Chinese would again raise the territorial issue in the joint commission and thus frustrate Moscow's desire to make tangible progress toward defusing the conflict.

EIGHT

The Moscow Conference

The Chinese Party Congress in April 1969 symbolized the radical isolationism that had marked Peking's policies since the onset of the Cultural Revolution. Contrary to tradition in the Communist movement, there was no foreign representation at the CCP Congress, and the policies propounded were those that had risen to the fore along with the Cultural Revolution forces, notably Lin Piao, now constitutionally enshrined as Mao's successor-designate. As the Chinese had retreated into isolationism since 1965, Moscow had been intent on seizing the opportunity to recover ground in the international Communist movement that it had lost to Peking's ideological challenge in the first half of the decade. It was for this purpose that the post-Khrushchev leadership had fashioned its line of "united action" in support of the Vietnamese comrades. Initially, in order to demonstrate its *bona fides* in offering to put aside the ideological rivalry in the interest of fraternal unity, Moscow had shelved Khrushchev's project for holding a new international party conference that, in effect, would have isolated China from what the Soviets portrayed as the mainstream of the international movement. By late 1966, however, Peking's intransigent rejection of the Soviet line of unity prompted Moscow to revive the project as an instrument in the political struggle with the Chinese. That project had been temporarily derailed by the invasion of Czechoslovakia but, after a difficult signaling and negotiating process,

Moscow was able to put its plans together again. The Sino-Soviet border crisis now lent the project even greater moment.

The International Communist Conference of June 1969 was a successor to the 1957 and 1960 conferences, which in turn were successors to the Comintern and the Cominform as instruments of international Communist consultation and coordination. The dissolution of the Cominform after Stalin's death was one of the steps toward destalinization, in this case a move by the new Kremlin leadership to remove the inheritance of Stalinist *diktat* from relations between Moscow and other Communist parties. It was the crisis in bloc relations resulting from destalinization, particularly from the impact of Khrushchev's major initiative at the 20th CPSU Congress in 1956, that occasioned the 1957 Moscow Conference, which issued a declaration of principles adopted by all the ruling parties except Yugoslavia. The Chinese having played a role in mediating the crisis in Soviet-East European Communist relations, the 1957 Conference marked a significant step toward according the Chinese a measure of authority in the determination of international Communist strategy and policy that had been the exclusive domain of Moscow as the arbiter of the movement's affairs. This was a step toward what has been likened to a "conciliar" phase of international Communist authority, in analogy to the ecclesiastical councils of Christendom which competed with the Papacy in determining doctrine binding on all the churches and believers.

It was the 1960 Conference, attended by 81 parties, that embodied the conciliar phase at its height. That conference spoke in the name of the international Communist movement as a whole, and issues affecting the interests of all parties, ruling and non-ruling, were brought for resolution in what was almost fully (the Yugoslavs were absent and were condemned) an ecumenical council. The Chinese now having mounted a serious challenge to Moscow's ideological and political authority on basic issues of international Communist strategy, the 1960 Conference registered a further erosion of Moscow's authority by bringing these issues for resolution to a tribunal constituted by the CPs of the world. There was, thus, a conciliar jurisdiction in which Moscow's influence was still strong but not absolute, and the interest of unity in the Communist movement required that the Kremlin accommodate itself to a locus of authority to which the Chinese and their allies could make an appeal.

The dissolution of the Comintern and its Cominform sequel, and their replacement by international conferences of a conciliar form, marked the end of democratic centralism as a principle governing the Communist movement as a unitary whole. The Comintern embodied a hierarchical organization of the Communist movement whose national "detachments" were expected to follow the line determined by Moscow as the central authority, an authority that became identified with Stalin personally. The 1957 Conference retained an element of democratic centralism in that Moscow was acknowledged as "the head" of the socialist camp and the Communist movement; but it had been mainly the Chinese for their own interests who had insisted on that formula, and thus this reflected the enhanced measure of Chinese influence arising from the destalinization crisis. The 1960 Conference removed that formula and acknowledged Moscow to be only the "universally recognized vanguard" of the Communist movement; the CPSU's experience was acknowledged to be of "fundamental significance for the entire international communist movement," but its authority as leader of the movement had become subject to Peking's challenge and to the jurisdiction of the other parties as a court of appeal to which both Moscow and Peking must now perforce have recourse.[1]

The weight of Moscow's traditional influence left a deep impress on the collective statement hammered out at the 1960 Conference, but the formulas of unity and accommodation composing that statement were too fragile to contain the polarizing forces in the Communist movement that Peking's assertive challenge and Moscow's forceful response had generated. The expectation that a conciliar phase of international Communist authority would be maintained was soon overwhelmed by accelerating polarizing pressures; and after the failure of CPSU-CCP talks and the publication of Peking's alternative "general line" for the Communist movement in the eventful summer of 1963, there could be little hope that another ecumenical conference could be arranged. The successor conference that Moscow began preparing in the last year of Khrushchev's rule was clearly to be a schismatic one, with Khrushchev resigned to a definitive fracture of the socialist camp that would remove the burden of the China question from Kremlin policy-making.[2] One aspect of Khrushchev's move on the China question was a partial disengagement from East Asia, particularly from the complication of the Indochina conflict. Accordingly, the post-Khrushchev leadership, in reversing his moves on China and related areas, combined an effort to get off a collision course with the Chinese, especially by shelving the schismatic conference project, with a move to reassert Soviet influence in Indochina and in East Asia generally, an effort made in the name of Communist unity. However, the logic of Sino-Soviet rivalry, and its implications for Chinese domestic politics, was such that this Soviet approach led to an ever more serious deterioration of relations and to a return to Moscow's agenda of Khrushchev's old plan for a schismatic conference.

Rather ironically, the new conference served both to dramatize the schism in the Communist movement and to provide a capstone to the line of unity of action in whose name the post-Khrushchev leadership had originally sought to avoid a schism. Thus, the official main topic of the June 1969 Conference and the title of its principal statement, "Tasks at the Present Stage of the Struggle Against Imperialism and United Action of the Communist and Workers' Parties and All Anti-Imperialist Forces,"[3] reflected Moscow's strategy of putting the focus on common interests in order to assemble the broadest possible array of parties as a backdrop against which to highlight Peking's isolation and intransigence. The success of this strategy can be measured by the appearance of 75 parties for the Moscow Conference, but the effects of the decade's polarizing forces were reflected in the absence not only of the Chinese and four of their Asian client parties but also of the other two ruling East Asian parties as well as the independent-minded

Japanese—all of whom had attended the ecumenical 1960 Conference. The Albanians joined their old enemies, the Yugoslavs, as absentees, and the Cubans sent an observer delegation. Hence barely a majority of the ruling parties were full participants.

The main document adopted at the 1969 Conference formally acknowledged existing reality by recognizing that "at this time" there could be "no leading center" in the Communist movement, and calling for "voluntary coordination" of the parties in conditions of "growing diversity." In effect, the Communist movement had passed from the Stalinist phase of democratic centralism during the Comintern era, through the conciliar phase of the ecumenical 1960 Conference, to what was now a loosely confederal framework in which individual parties took part or abstained at their discretion, and those participating in plenary conferences further exercised discretion by selective or complete acceptance or rejection of policy statements adopted by the majority. In keeping with this operative conception and in line with assurances during the preparations that there was to be no question of any excommunication, the main document explicitly endorsed the position taken by the Romanians that the socialist system comprised 14 countries, thus including China, Albania, and Yugoslavia. The 1960 statement, which contained a condemnation of Yugoslavia, had said that the socialist camp was composed of 12 countries, a formula serving to perpetuate the Cominform's excommunication of the Yugoslavs.[4]

Preparing the Conference

By accommodating itself to polycentrist trends and seeking a wide basis of consent, Moscow had succeeded in significant measure in containing the centrifugal forces generated by the Czechoslovak crisis, one that threatened to sever the reformist wing of the Communist movement as the radical Asian wing had earlier become separated from Moscow's mainstream. It was its management of post-Czechoslovak affairs that was a key to Moscow's success in mounting a conference with sufficiently broad representation to serve its purpose on the China question. Moscow had taken a low posture on the conference project during the period immediately after the invasion of Czechoslovakia, seeking a breathing space for arranging "normalization" of the situation in Czechoslovakia and for weathering the storm of protest aroused in parts of the Communist movement. Following an announcement on 1 October 1968 of the postponement of the conference, Moscow had signaled authoritatively that only the timing, not the desirability, of a conference was in question, the message being that the project would be carried through once the Czechoslovak complications had been surmounted. A treaty signed on 16 October regularizing the status of Soviet forces in Czechoslovakia served as a watershed in Moscow's approach to the conference question. Where Moscow had previously avoided raising the Czechoslovak and conference questions in the same context, now the Soviets began to try to come to terms with these issues and to settle the dust in the Communist movement raised by the invasion. Two articles in *Pravda* by foreign Communist leaders acting as Soviet proxies illustrated this approach. Before the signing of the treaty legalizing the presence of Soviet troops in Czechoslovakia, an article in the 9 October *Pravda* by the Syrian CP leader Bakdash defended the invasion against its critics in the Communist movement, but the conference question was evaded.[5] In contrast, a week after the treaty was signed, an article in the 23 October issue by the Lebanese Party's general secretary not only justified the invasion but went on to cite the treaty in an appeal for convening the international conference: "The conclusion of the treaty creates a good basis for a reconciliation of the viewpoints in the world communist movement. The Lebanese Communist Party considers, as before, that the convocation of a world conference of fraternal parties is an urgent necessity."[6]

In keeping with this characteristic use of proxy spokesmen, much of the momentum in the renewed campaign to stage a conference derived from foreign party statements, several of which lent the tone of urgency marking the Lebanese Communist's appeal. Moscow chose a meeting with a high-level delegation from the French CP, long a loyalist supporter but a critic of the invasion of Czechoslovakia, to give its first indication of the timing of the conference; a joint communiqué urged that efforts be made to hold the meeting in the "near future,"[7] a prescription echoed by Brezhnev and a parade of other Communist leaders attending the Polish Party Congress in mid-November.

The use of the French CP was symbolically valuable in Moscow's move to surmount the effects of Czechoslovakia on the conference project, but a more significant test was presented by the more independent Italian Communists. That the latter were not easily won over was reflected in a joint communiqué on talks in Moscow from 13 to 15 November between the Italian CP's Enrico Berlinguer and Brezhnev's deputy, Andrey Kirilenko.[8] In contrast to the Soviet-French communiqué, this one omitted any reference to the Czechoslovak issue, and the indecisive result of the talks as regards the conference question was reflected in the communiqué's bare assertion that the delegations had "exchanged" views on the subject. At the same time, the brunt of the Soviet bloc's counterattacks on the Italian and other West European Communist critics of the invasion was being borne by the Poles and East Germans, strong proponents of a centralist line in Communist affairs. At the Polish Congress, Gomulka chided the CPs of "developed capitalist countries" for seeking to impose their strategy on the bloc parties. He conceded that it was proper for them to utilize parliaments and other bourgeois institutions to further their objectives—"the roads to socialism can and should vary"—but he called it impermissible for them to demand that ruling parties "should adapt their political line to the policy pursued by the Western Communist parties at a given moment." This reflected the orthodox parties' resentment toward critics who were objecting to the continuing practice of Stalinist authoritarianism in the East bloc countries.

If the Poles were arguing the superiority of authoritarianism over liberal democracy, the East Germans, nervously mindful of the allures of the other part of Germany, were not even disposed to grant the validity of the Western CPs' policies in their own settings. At an East German Party Plenum held from 22 to 25 October,

Politburo member Kurt Hager bitterly reviled "revisionists" who promoted "dangerous illusions about bourgeois democracy by overemphasizing certain successes" in a bourgeois institution like a parliament. Joining issue with those who questioned the universal significance of the Bolshevik Revolution by viewing it as a "nationally limited event in a backward country," Hager rebutted the argument advanced in the Italian Party's *Rinascita* that "the alternative between social democratic opportunism and Stalinist dogmatism" can be by-passed by following "the Italian road to socialism"—that is, a road stressing democratization of socialism, as the Czechoslovak reformers had dreamed. Rejecting this alternative, Hager argued that the real issue was "the abandonment of the truth about the general validity of Leninism, about the world historical role of the October Revolution and the Soviet Union."[9] The *Rinascita* article, by Italian Party official Achille Occhetto in the 6 September issue, had cited Marx as expecting that socialist revolution would take place in advanced capitalist countries so as to assure it would "coincide with an immediate expansion of democracy," as well as Lenin's judgment that the failure of revolution in the West and Russia's economic and cultural "backwardness" necessitated "absolute centralization" of power in the Soviet Union.

An important step toward convening the conference was taken shortly after the Polish Congress, where Brezhnev had enunciated the doctrine of limited sovereignty and led the drive for a conference to be held in the near future. A meeting of the preparatory commission on 18–21 November arranged a new timetable for the conference, now scheduled for May 1969 and to be preceded by another preparatory commission meeting on 17 March to examine the draft documents and set an exact date. There was an impressive total of 67 parties present, including Romania and the Western CPs critical of the invasion of Czechoslovakia. The decision to convene the conference in May drew an expression of "deep satisfaction" from a *Pravda* editorial on 25 November 1968, the date the conference was originally scheduled to open. Reflecting Moscow's growing sense of confidence, *Pravda* injected a significant ideological note by promising a rebuff not only to imperialist adversaries but also the "rightwing and leftwing opportunists and splitters" damaging the Communist movement.[10] Clearly, there was still considerable potential for friction between Moscow and the parties voicing autonomist sentiments, as shown in Romanian leader Ceausescu's forceful reaffirmation of independence in speeches on 28 and 29 November in which he drew on strongly nationalistic themes from Romanian history. Declaring that "it is our duty to develop socialism in Romania," he defiantly added: "If this be nationalism, yes, comrades, we are nationalists; we want communism in Romania." He anticipated another issue in the conference preparations by declaring that "there is no need for a leading center" in the Communist movement and reiterated Romania's determination to have good relations with "all" parties.[11]

It was in the course of what clearly were difficult preparations of the main document for the conference that the strains and divergent approaches of the various parties were given sharper definition. As it happened, the March 1969 meeting of the preparatory commission coincided with the semicentenary anniversary of the founding of the Comintern, an ideal occasion for venting the divergent centralist and autonomist trends in the current situation. Given the divergences being aired at this time, it was not surprising that the preparatory commission in March again set back the date of the conference, saying that another and final session of the commission would meet in May and that the conference itself would open on 5 June. In the meantime, the two East European countries that had stood up for the Czechoslovak reformers, Romania and Yugoslavia, were intent on putting the Comintern in historical perspective. Romania's *Scinteia* carried a Comintern anniversary article on 11 March which acknowledged the body's early contributions but stressed that "new prospects" subsequently emerged which "demonstrated more and more the anachronism and incompatibility" of the international Communist movement's being "led from a center." Drawing the implications for Romania's position on the forthcoming conference, *Scinteia* granted that conferences "may be useful" but warned that they must not make "binding decisions" or "pass verdicts" on individual parties.[12] The *Scinteia* article coincided with the opening of the Ninth Congress of the Yugoslav League of Communists, a congress timed for the 50th anniversary of the Yugoslav Party. Romania alone among the East European countries was represented at the congress. Tito's opening address reviewed the party's turbulent history, relating a story of development arising from the country's "historical and national characteristics" in the face of Soviet interference dating back to the Comintern years. Tito used the occasion not only to air his grievances against the Comintern and Stalin personally but also to contrast Yugoslavia's "democratic concept of socialist society" with the "bureaucratic" socialism under Stalin. He also made the point that time had vindicated the Yugoslav Party's defiance of the Cominform's charges of 1948 and had "eloquently refuted" the "arbitrary and crude" anti-Yugoslav attacks made at the 1960 Moscow Conference.[13]

Meanwhile, Moscow was drawing different lessons from the experience of the Comintern, whose anniversary served as a major instrument in Soviet moves to prepare the conference in the most propitious possible atmosphere. A Comintern anniversary meeting held with international representation in Moscow on 25–26 March was used to counter the criticism by the independent-minded parties and to defend the relevance of the Comintern's ideological legacy for the present situation, particularly in the form of "proletarian internationalism" as a counter to autonomist tendencies. But it was of equal significance that Moscow used the occasion to signal a flexible approach to conference preparations and to reassure the parties that no attempt to revive Comintern organizational forms was envisioned. It was important for Moscow's authority to claim a line of continuity running from the Comintern to the conference now being planned, but in the circumstances it was also prudent to acknowledge mistakes in the past as a way of reassuring the critics that their concerns were being given a hearing. In fact, Moscow's move to meet part way the objections of the Romanians and other critics went too far for some of the more centralist-minded elements in the Soviet bloc.

The major Soviet figures at the Comintern anniversary meeting were senior ideologue Suslov and his deputy Ponomarev, the party secretary in charge of international Communist relations. In delivering the opening address at the meeting, Suslov[14] linked the Comintern's legacy with the present situation by asserting that the Marxist-Leninist parties, "following the Comintern's traditions, are waging a constant struggle against rightwing and 'leftwing' opportunitism, bourgeois and petit-bourgeois nationalism, and all departures from Marxism." But, while making use of the Comintern as a precedent for vigorous defense of orthodoxy in Communist affairs, Suslov at the same time acknowledged "errors" in the Comintern's activities, among these being the failure at times to take proper account of different national conditions, the thesis promulgated for a time that social democracy was the main danger to the Communist movement, and the "adverse effects" of Stalin's "personality cult."[15] He further conceded that the Communist movement had outgrown the organizational form in which its leadership was exercised from one center, as it had during the Comintern era when the situation demanded "obligatory international discipline" in accordance with the principles of democratic centralism governing the movement as a whole. Noting that there were forms of international Communist ties other than those characteristic of the Comintern, Suslov cited international conferences such as the one being prepared as a major form befitting current conditions.

Ponomarev's more extensive report, "The Historical Significance of the Comintern,"[16] also acknowledged the Comintern's mistakes but went to considerable length to stress the primacy of proletarian internationalism and to counter attempts to belittle the Comintern's historical role as a way of attacking its internationalist principles. He was concerned to defend Leninism against those who would divorce it from Marxism or would impart a limited national character to Leninism. But, while thus defending the tradition of Soviet ideological authority, Ponomarev's report elaborated on the concessions of past error made by Suslov in an effort to find common ground with the critics. Thus, in addition to the mistaken thesis on social democracy as the main danger, Ponomarev cited the Comintern Sixth Congress's denial of the anti-imperialist potential of the national bourgeoisie as an error; and regarding the adverse effects of Stalin's personality cult, Ponomarev discussed the related deviations from "Leninist norms of party life," chiefly the improper treatment of cadres, a deviation that could also be ascribed to Mao's Cultural Revolution. The point was that it was not Leninism but deviations from those principles that marred the Comintern's otherwise positive record. Moreover, Ponomarev's report repeatedly cited the Comintern Seventh Congress in 1935 as an example of the organization's ability to rectify deviations. That congress had sanctioned the popular front policy, a historical memory particularly suited to Moscow's current effort to assemble a wide range of parties in the name of united action. It also provided a useful background for an attack on the Chinese as a threat to Communist unity in the present situation.

Ponomarev made use of the thesis that there was now a "deepening and exacerbation of the class struggle in the international arena" to underscore the urgency of proletarian internationalism and of a "resolute repulse of rightwing and 'leftwing' revisionists and splitters and of nationalism." Claiming that international conferences represented a form of coordination of views of "equal and independent" parties in conditions in which there was not a single organizational center of authority, Ponomarev looked forward to the scheduled conference as an opportunity for "a collective Marxist-Leninist analysis of the present situation" and for the elaboration of a "platform of common actions."

As it had during the period of preparations in 1968 for the conference then scheduled for November that year, Moscow now again set forth its case on the China problem, using the authoritative vehicle of editorial articles in the theoretical journal *Kommunist* to convey its analysis of the Chinese heresy. The first in the new series dealt with the Chinese internal situation, especially in view of the anticipated CCP 9th Congress.[17] According to the introduction to the series, these articles presented a "principled Marxist-Leninist evaluation" of the situation in China in order to assist the "genuine" Communists of that country to understand what was taking place there and to assist their struggle to restore the policies set forth at the 1956 CCP Congress. The burden of the first article's analysis of current trends in China lay in the argument that a new stage had arrived in which the anticipated congress would be virtually a constituent congress for a new political organization replacing the Marxist-Leninist party that the Cultural Revolution had decimated. The implication of such an analysis was that Moscow had not only the right but the duty to seek to isolate the Maoist leadership, which was in the process of erasing a detachment of the Communist movement.

The line that a new organization was replacing China's Marxist-Leninist party suggested the possibility that Moscow might be tempted to promote a CCP-in-exile. Whether or not such a notion was seriously considered in the Kremlin, the length to which Moscow was now prepared to go to challenge Mao's authority was strikingly demonstrated late in March—the month of the initial border clashes—by the resurfacing of Wang Ming after many years of oblivion and exile in the Soviet Union. Having been trained in the Soviet Union before becoming the CCP's general secretary in the 1930s, then becoming the Comintern's representative after Mao assumed control of the party, Wang was a prime example of what Moscow had been calling the "Communist-internationalists" in China, meaning those amenable to Soviet influence. It was probably to avoid appearing too directly to be promoting Wang's renewed challenge to Mao at this time that Moscow used the pro-Soviet Canadian CP's paper *Canadian Tribune* as the original publication source for a long article by Wang virulently reviling Mao's Cultural Revolution. Consistent with the *Kommunist* analysis, Wang concluded that, at the impending CCP Congress, a "new Maoist anti-Marxist, anti-Leninist reactionary party, 'Communist' in name but anti-Communist in substance," would be formed.

Wang's article was given worldwide dissemination in Soviet media, including extensive excerpts broadcast in installments to the Chinese audience and republication of the text as a pamphlet by Moscow.

Another indication of Moscow's hardening line on the China question in the period of preparations for the conference was the intransigent treatment of Peking's foreign policy in the next *Kommunist* editorial article, this one having been prepared after the border clashes and coinciding with the decision to resurface Wang Ming.[18] The article reflected Moscow's awareness of the signals appearing from Peking since the previous autumn that new directions were emerging in Chinese foreign policy. Thus *Kommunist* took note of Chou's declaration in September 1968 that the socialist camp no longer existed, and it also discerned signs of a new approach by Peking on the Vietnam question, a change ascribed by *Kommunist* to a Chinese fear of being left out of a settlement and especially to a Chinese calculation on rebuilding relations with the capitalist states by paying an appropriate price in Indochina. This perception by Moscow did, in fact, anticipate the more flexible approach by Peking to the Vietnam question, one in which the Chinese were willing to encourage or at least to avoid undermining moves toward a settlement in the broader interests of Chinese policy in East Asia and in the Sino-Soviet conflict. *Kommunist's* analysis also cited the significance of Peking's November 1968 overture to the United States, a move that was seen as related to Peking's hopes that the newly elected President Nixon's campaign pledge to reduce U.S. commitments would be realized first in Asia. In all of this, *Kommunist* saw the recent border clashes as a central link in Peking's anti-Soviet policy.

Moscow's tough approach to the China problem was signaled in the editorial article's stern warning against any underestimation of the danger posed by Mao's policy to the Chinese people and the Communist movement. *Kommunist* drew the lesson that any attempts at compromising with Mao's China were in vain, and thus "Marxist-Leninists must not relax their efforts in the struggle against the theory and practice of bellicose Maoism." Notwithstanding Moscow's assurances during the conference preparations that no excommunication was contemplated, the signals from Moscow during this period strongly suggested an intent to make the Chinese the prime target of the conference. Neutralist sentiment within Communist ranks regarding the Sino-Soviet conflict thus represented an obstacle to Moscow's goals. Bucharest was being meticulously neutral on the border dispute, as registered in the party paper *Scinteia's* practice of publishing paired reports, one datelined Moscow and the other Peking, setting forth each side's respective versions of the border fighting.[19] It was that kind of neutral reporting that drew from Moscow a complaint about "objectivist" coverage, a curious notion suggesting that to seek to be objective was a pose and that the truth was necessarily partisan. An even more revealing measure of Moscow's acute concern over neutralist views in the Communist movement at this critical juncture could be found in a bitter rejoinder in *Pravda* on 13 April to an article in the Italian CP's *L'Unita* four days earlier which attached reservations to a statement of support for the Soviets against Peking's territorial claims. The *L'Unita* article, by Giuseppe Boffa, firmly rejected Peking's claims as "absolutely unjustified," but it also observed that Moscow had weakened its position in the eyes of international left-wing opinion by stressing nationalistic propaganda and by its intervention in Czechoslovakia. Boffa also invoked the views of another party by approvingly citing Spanish CP leader Carrillo to the effect that "unconditional defense" of the USSR was a valid course for the Communist movement when that country was isolated and encircled, but that, when there were conflicts between Communist countries, the problem became more complex: "The old conditioned reflexes of the time when the Soviet Union was the only socialist country are no longer enough."

The 13 April *Pravda* response, by I. Ivanov,[20] professed a preference for answering Boffa "privately" but argued that all who supported Communist unity would "understand" that the airing of Boffa's views in public dictated a published response. In fact, however, Moscow had avoided replying since August to *L'Unita's* outspoken criticism of Soviet behavior on the Czechoslovak question, some of it by Boffa himself and some of it directly rebutting *Pravda* articles defending the invasion. The meaning of the Ivanov article was that, in this period of border tensions and conference preparations, Moscow was intent on putting the parties under pressure to stand up and be counted on the Soviet side in the conflict with China. Thus, the article took the position that the Soviet Union was defending interests broader than its own national interests in the border crisis. In addition, the article linked the China and Czechoslovak questions more closely than in previous Soviet commentary, invoking proletarian internationalism to demand that Communists "close ranks, compactly and firmly," in the face of danger, and citing the Czechoslovak situation and the border clashes as cases of such danger. The effect of the border crisis in hardening Moscow's line in Communist affairs had also been evidenced in a tough *Pravda* article on 7 April, the date of the Romanian foreign minister's arrival in Moscow. That article, by E. Bragamov,[21] warned in strong terms of the danger of nationalism in the Communist movement, thus using the code word for that deviation from Moscow's line of which the Romanians were notorious practitioners. According to Bragamov, the present conditions, in which the Chinese were creating a situation where the struggle against nationalism was "much more complex," required a solid front against Peking's "chauvinistic attacks."

These warning signals of a hardening of Moscow's line were strongly reinforced by a *Kommunist* editorial in May[22] in anticipation of the International Conference. The editorial called for "an irreconcilable struggle against all revisionist distortions from the right and 'left' " and expressed Moscow's concern over the "purity" of Communist doctrine, thus indicating an intent to use the conference to enhance Moscow's ideological authority. Significantly, the editorial reasserted the doctrine of limited sovereignty in defending the intervention in Czechoslovakia, and delivered a strong attack on China for the "particularly harmful" effect of its "great-power, chauvinistic policy." The same issue of *Kommunist* also carried an article putting into Soviet perspective the deterioration in relations with China in the decade since the 1960 Conference sought futilely to accommodate the burgeoning conflict.[23] In noting that the Chinese had tried to impose their views on the 1960 Conference after having initiated their ideological challenge to Moscow earlier that year, *Kommunist* recalled with satisfaction that the conference had rejected Chinese attempts to " cast doubt on the universally recognized authority of Lenin's party." One purpose of this review of a decade of Sino-Soviet relations was to defend the Brezhnev leadership's China policy and to demonstrate that the Chinese had demanded a complete Soviet capitulation to Peking's ideological authority as the price for accepting Moscow's proposals to normalize relations after the ouster of Khrushchev. The article also dealt at length with the border dispute, paying tribute to the Kremlin's "exceptional self-control and calm" and stressing its proposal to resume the 1964 consultations.

It was to be expected that Soviet pressures for closed ranks in the Communist movement would complicate the drafting of the conference's main document, and this was reflected in the labors of the final meeting of the preparatory commission, held from 23 to 30 May. That lengthy session encountered difficulties in achieving consensus; the final communiqué said noncommittally that the delegations "set forth the opinions" of their respective parties on the main document and agreed to "refer the draft, with the amendments included in it, as a basis for the work" of the conference.[24] It was a matter of particular concern to Moscow and its loyalist allies to mute the Czechoslovak issue, the cause of the earlier postponement of the conference and the potential source of a further schism that would undercut Moscow's campaign to isolate the Chinese. This purpose had been served the previous month when Alexander Dubcek was replaced as Czechoslovak party chief by Gustav Husak, one of the most effective proponents of compliance with the "normalization" prescribed by Moscow. The ouster of Dubcek from party leadership came in the wake of anti-Soviet rioting in Czechoslovakia in late March, an inflammatory development in any situation but the more so in coming amid the tensions of the Sino-Soviet border fighting. This leadership switch facilitated Moscow's labors in the final phase of conference preparations, as when the Czechoslovak representative at the May preparatory commission meeting, the conservative Vasil Bilak, returned home in the middle of the session to report to a party plenum. The plenum issued a statement expressing "resolute disagreement" with the view that Czechoslovakia should be made a subject of discussion at the conference "in any form." The pro-Soviet parties were thus armed with an express appeal from the Czechoslovak Party itself not to subject its affairs to judgment at the conference.

THE CONFERENCE CONVENED

That such a broad array as 75 parties could assemble in Moscow on 5 June 1969 to hold a new International Party Conference lent a measure of substance to the appearance desired by the Kremlin of the Communist mainstream still flowing through Moscow. This mainstream had now arrived somewhere along the way between the old monolithism and a thoroughgoing polycentrism; Moscow's purpose was to arrest the flow by using its influence and power to shape the conference into an instrument of cohesion and common purpose. To make a success of this project, however, the Soviets and their loyalist clients were constrained to accommodate themselves to the forces of independence and dissent. The results and implications of the conference must be assessed in terms of Moscow's success in managing this accommodation in order to realize the central purpose of isolating China.

Moscow had exhibited awareness early in the conference preparations that the conciliar phase of the Communist movement had foundered on schism and that it would not be possible to push through a set of ideological principles binding the Communist movement as a whole. In his address to the initial Budapest consultative meeting in February 1968, Suslov had signaled this awareness by tactically equivocating on the nature of the basic document the projected conference would adopt. While proclaiming Moscow's fidelity to the "general line" of the international movement formulated in the documents of the 1957 and 1960 conferences, and asserting that experience had confirmed the validity of that general line, he indicated that Moscow did not expect a reaffirmation of "every letter" of the documents of the preceding conferences.[25] Actually, Moscow's hopes at that time, before the disarray attending the invasion of Czechoslovakia, had probably been higher regarding the prospects of establishing a firmer line of continuity with the 1960 Conference. In the event, the main document adopted in 1969[26] studiously avoided the question of the earlier documents' status, cautiously affirming only that the past decade of events had borne out "the Marxist-Leninist assessment" of the nature of the present era. In contrast, the 1960 statement had explicitly established its continuity with the 1957 document as having been confirmed by the events of the intervening three years.

Also, in keeping with the constraints imposed by the need for wide representation at the conference, the China question was excluded from the main document in recognition of the tenacity of Romania's and others' opposition to any appearance of an excommunication of any party. But China was the reality that lay behind the whole exercise, from Moscow's point of view, and, in the complex interplay of signaling during the conference preparations, the Soviets had made clear their determination to use the conference as an instrument in their conflict with the Chinese. Accordingly, the conference was turned into a forum for condemning the Peking heretics, with Brezhnev's ringing denunciation of Mao's China accompanied by a parade of loyalists voicing their support. In this well-orchestrated campaign, the opening shot in the anti-Chinese fusillade was fired on the second day by the obscure Paraguayan delegate. This choice of speaker and timing—the first day having heard several pro-Moscow delegates without a mention of China—was presumably contrived to avoid the blatant appearance of a conspiracy by Moscow to circumvent demands by the Romanians and others that no absent party be censured. Immediately following the Paraguayan's address, Ceausescu took the floor with an appeal against this behavior and a warning that the conference could be jeopardized if other delegations followed suit. His appeal, reminiscent of Chou En-lai's rebuke to Moscow for Khrushchev's attack on the Albanians at the 22nd CPSU Congress in 1961, proved to be in vain as a total of 50 parties at the conference joined in the anti-China chorus.[27]

After Ceausescu's intercession, the next scheduled speaker after the Paraguayan was Poland's Gomulka, whose task was to present the pro-Soviet case against China from the standpoint of bloc interests. Gomulka advanced the thesis that internationalism constituted the basic criterion for judging the correctness of each party's policy, and in this context he accused the Chinese of having betrayed the principles of internationalism. Such a thesis, and his complementary identification of nationalism as the chief ideological source of the threat to Communist unity, put the weight heavily on the centralist side of what Romania sought to be a balance between collective responsibilities and individual party rights. Similarly, Gomulka asserted that the Soviet Union remained the "main center" of the forces of socialism, explaining that the security of all socialist states, including China, depended on the Soviet Union's strength. Though China rather than Romania was named, the latter was a primary audience for Gomulka's message.

The main burden of presenting the pro-Soviet loyalist case, including a rebuttal of Ceausescu's appeal, fell on the willing shoulders of Bulgaria's Zhivkov, who expressed the most atavistic sentiments in making a strong plea for Soviet leadership of the international Communist movement. Zhivkov took pride in his party's role as Moscow's spearpoint, beginning with the use of the Bulgarian Party Congress in November 1966 to relaunch the campaign for an international conference. He cited his countryman Georgi Dimitrov, once the Comintern's Secretary General, as having defined one's attitude toward Moscow as the "touchstone" of proletarian internationalism. Zhivkov added that the significance of that postulate was "even greater today" in that the emergence of a world socialist system, in addition to its positive aspects, contained opportunities for the appearance of nationalistic trends in socialist countries; moreover, "some people" in the Communist movement at the present time attempted to undermine proletarian internationalism behind the mask of defending the equality and independence of the individual parties. The logic of Zhivkov's argument, and certainly its spirit, suggested that in the present situation the need for subservience to Soviet authority was even greater than when the Soviet Union was an isolated and encircled bastion of socialism.

In responding to Ceausescu's intercession, Zhivkov stressed that it was incumbent on the parties to deal with the China problem because some (unnamed) CP leaders were regarding it as a bilateral matter between Moscow and Peking; he emphasized in strong terms that the China question presented " a profound matter of principle" affecting the fate of the entire revolutionary movement. He added that Moscow's supporters would have heeded the plea to avoid raising the China question if the Chinese had not rejected the invitation to the conference "so arrogantly" and "intensified" their struggle against the international Communist movement.

The other of Moscow's allies given the most extensive anti-Chinese script at the conference was the Mongolian leader Tsedenbal, speaking in behalf of a party with special grievances against Peking and as a partner of the Soviets in the border confrontation with China. Tsedenbal used the China question to assert Moscow's primacy in the Communist movement, charging that the Chinese were directing their attack "against the CPSU as the most steadfast and experienced detachment of the communist movement" and "against the Soviet Union as the main and powerful support of the socialist community." Much like Zhivkov, he served the Kremlin's cause by stressing that the Sino-Soviet border conflict could not be regarded as a matter concerning only Moscow and Peking but directly concerned the fundamental interests of the whole Communist movement. Tsedenbal also used the occasion to air his party's grievances against the Chinese, recalling Mao's 1964 interview with the Japanese as indicating Peking's desire to annex Outer Mongolia and citing the "oppression and humiliation" of the Mongols, Kazakhs, Tibetans, Uighurs, and other minority nationalities in China as a measure of what would become of the people of the Mongolian People's Republic if Mao's desire were realized.

And so the parade of pro-Soviet loyalists vented their ire at the Chinese and sought to demonstrate a ground swell of support in the Communist movement for Moscow's cause in the Sino-Soviet crisis. Charges of Chinese provocations on the border figured prominently in the litany of denunciations of Peking that rang out from the conference rostrum. A key to this bandwagon was revealed by the U.S. Party's delegate, Gus Hall, who stressed in his address that no anti-imperialist, "and certainly no communist," could take a neutralist position toward Peking's anti-Soviet behavior. There may have been an element of deception behind the rush of anti-Chinese attacks that Ceausescu's appeal failed to stem, and his own surprise at the turn of developments was indicated in his remark after the Paraguayan's initial sally that the Romanians had welcomed the fact that censure of other parties had been renounced during the conference preparations. It may have been with this in mind that Hungary's Kadar, the one among Moscow's close allies with the best repute for moderation, used a press conference on 14 June—three days before the conclave ended—to explain the catalytic effect of the border clashes. Kadar said that those clashes, which he duly blamed on the Chinese, had "brought about a change in the position of some parties at the conference, in the sense that they began to display an even more serious attitude" toward the Sino-Soviet conflict. This observation was picked up in the Soviet media.

If the China question dominated the conference proceedings, the Czechoslovak issue stood in significant contrast as a carefully submerged matter, and both these results testified to the effectiveness of Moscow's management of the conference. Like the China question, Czechoslovakia emerged as an issue on the second day of the proceedings when the outspoken Australian delegate condemned the invasion, but otherwise there was little resemblance in the way the majority of delegates treated these issues. A mere seven other parties raised the Czechoslovak issue in a critical vein, and six delegations specifically defended the intervention. The first of the latter was the El Salvador delegate, who immediately responded to the Australian's criticism by saying that his party—"which is directly fighting against imperialism and from its own experience knows the insidious nature of its methods"—regarded the intervention as "timely and necessary," but he added that the subject was "irrelevant" in any case because the Czechoslovak representative's statement to the preparatory commission had stressed that discussion of the Czechoslovak question conflicted with the "normalization" of the situation in that country. By ruling the matter out of bounds, the El Salvador delegate conveyed the essence of Moscow's strategy of muting the Czechoslovak issue and removing it from any jurisdiction of the broader Communist movement.

Among the half dozen delegates defending the intervention in Czechoslovakia, the role of chief apologist was assigned, understandably, to Kadar, whose credentials included not only his reputation as being the leader among the invading countries showing the most independence but also his position as the leader of a country which itself had undergone the experience of invasion by its Warsaw Pact allies. Kadar cited Hungary's 1956 experience as illustrating the way in which the imperialists sought to undermine socialist systems, and he explained that the "international forces" intervening to assist Hungary had in mind that the defense of the workers' power and socialist gains in one country was "the common cause and interest of every socialist country." It was against the background of the Hungarian experience that he justified the intervention in Czechoslovakia, though he evaded a direct reference to the invasion as such. From all of this, Kadar in effect summed up the experience of the Hungarian and Czechoslovak crises with a variation of the doctrine of limited sovereignty: "As shown by experience, when the destiny of a socialist country is at stake everyone is concerned . . . and when the leaders of a socialist country make proletarian internationalism independent of the interests of their state, then they can only do harm to the common cause as well as to the interests of their own country."

The only other bloc leader to broach the Czechoslovak question explicitly was the new Czechoslovak Party chief, Husak, who diagnosed the origins of the crisis as lying in the serious weakening of the CP's leading role during the liberalization movement. He described the April and May 1969 Plenums at which he consolidated his power as turning points, and he scornfully dismissed those parties at the conference which, "with a very superficial knowledge of our affairs," had been drawing "hasty" conclusions about the Czechoslovak situation. He added that he realized some of these parties were under "the pressure of bourgeois and petit bourgeois propaganda." So much for the right of the reformist parties to raise the issue of Soviet interventionism posed by the invasion.

THE SOVIET POSITION

The familiar technique of using proxy communication was much in evidence in Moscow's management of the conference, especially in conveying the outer reaches of the pro-Soviet lines; but the most direct measure of Soviet expectations must be found in the address Brezhnev delivered on the third day of the proceedings.[28] As in the case of the conference as a whole, Brezhnev's address was dominated by the China question, underscoring the centrality of that question in Moscow's calculations in deciding to revive Khrushchev's schismatic project. In what may have seemed a paradoxical element in an affair officially billed as a rallying of anti-imperialist forces under the banner of Communist unity of action, Brezhnev's statement was permeated with the language of détente, including a notably strong signal to the United States that this language was to be taken seriously. The paradox diminishes, however, if the origins of the China problem are recalled; after all, it was Peking's antagonizing reappraisal of the meaning of the Sino-Soviet relationship in view of Moscow's détente policy toward the West that gave rise to polarizing forces in the Communist movement. Moreover, the effect of that polarization in keeping the radical Asian wing of the Communist movement away from the conference gave Moscow nearly free rein to preach the virtues of détente, virtues that were questioned by only a fragment of the

Latin American Communist movement upholding the banner of militant *fidelisma*.

In keeping with the generally muted treatment of the Czechoslovak question at the conference, Brezhnev was almost vaporous in his formulation of the issues involved. Indeed, he made a point of chiding "imperialist" propaganda for circulating a theory of limited sovereignty, as if to deny any substance to what was subsumed under that name. He did, however, identify the source of right opportunism as deviation from the "general laws" of socialism, such as the dictatorship of the proletariat and the leading role of the CP, the key issues of the Czechoslovak crisis. But the Brezhnev doctrine he had so forcefully asserted in Warsaw the previous November was here watered down to the formulation that "any weakening of socialist positions in the world is bound to reflect negatively on the positions of all the communist parties," a formulation leaving unspoken the crucial operative assumption of an interventionist right to offset any such weakening of socialism in a fraternal country.

The intervention Brezhnev was concerned to justify at this forum, in this case a political and ideological intervention, was the use of the conference to wage Moscow's conflict with Peking. In introducing his lengthy and sharply polemical discourse on the China question with a discussion of the struggle against opportunism and nationalism in the Communist movement, he acknowledged that this struggle should fall primarily within the domain of the party concerned, but he argued that, when this struggle was abandoned in one segment of the movement, its effects were felt by the movement as a whole. Here he pointed to the CCP leadership's position as "a striking example of the harm that can be done to the common cause of communists by departing from Marxism-Leninism and breaking with internationalism." He thus in effect rejected Ceausescu's appeal for the conference to avoid passing judgment on other parties. He did remark, however, in an implicit admission that the Romanians and other neutralists in the Sino-Soviet dispute may have been deceived during the conference preparations, that until "recently" the Soviets had had no intention of broaching the China question. He explained that recent events, "particularly the nature of the decisions" taken by the CCP Congress that April, had compelled the Soviets to deal with the "new situation" that was having "a grave negative influence" on the international environment. Taking the recent CCP Congress as marking a "new stage" in the evolution of Maoism, he noted in particular that the Chinese Party's new constitution adopted at the congress defined Mao's thought as Marxism-Leninism of the present era. Brezhnev thus zeroed in on that Chinese claim to ideological authority that most affronted the other parties in the Communist movement, ranging from the pro-Soviet loyalists through the independents like the Italians to such absentees as the North Vietnamese.

It was a well-chosen tactical ploy to cite the CCP 9th Congress, and specifically its claim in behalf of Maoist universalism, as the reason for raising the China question at the conference, but the basic underlying motive related to another aspect of recent events that also figured in Brezhnev's discourse, namely the eruption of fighting on the border. In a revealing passage going to the heart of the matter, Brezhnev emphasized that it was "doubly important" to deal with the China problem because "a section of progressive world opinion still believes that the present Chinese leadership has revolutionary aspirations" and struggles against imperialism. This reflected Moscow's acute concern and resentment over neutralist positions taken by some parties toward the Sino-Soviet conflict, a resentment that had boiled up on the pages of *Pravda* in April in the bitter rejoinder to the Italian Party organ.[29] Much as Suslov at the original Budapest consultative meeting had used Peking's "provocative" rejection of the invitation to that gathering, Brezhnev called attention to the "insults" showered on "all" the conference participants by Peking in its reply to their invitation. Moreover, he cleverly ticked off a series of parties, including the Italian CP, as among the "overwhelming" majority of CPs the Chinese had branded as revisionist. And, while thus playing up Peking's isolation and intransigence in Communist affairs, Brezhnev was also intent on portraying China as a threat to international stability, which meant a threat to conditions desired by many of the parties as being propitious for their own interests. In addition to imputing to Peking great-power aspirations and territorial claims to other countries, Brezhnev used a favorite polemical line from the Khrushchev era by depicting Mao's China as dangerously irresponsible on matters of war and peace. Here he quoted Mao as having spoken at the 1957 Conference with "startling lightness and cynicism" about the prospect of the destruction of half of mankind in a nuclear war. Updating this line to cover the current border confrontation, Brezhnev declared that Peking's combining of "political adventurism with the atmosphere of sustained war hysteria" introduced new elements into the international situation, "elements which we have no right to ignore." He thus set the stage for presenting the Soviet case on the border conflict.[30]

Brezhnev's treatment of the question of détente was complementary to his discourse on China as a factor of dangerous instability in the international system. The end of the 1960s, a period of upheaval and violent disruptions of the social and institutional fabric of the Western world, was a time that might have presented inviting opportunities for the Communist movement, but the heir of Lenin's Bolsheviks chose to offer a soberly realistic appraisal of the strength and resilience of "imperialism." It was this appraisal that underlay his strong affirmation of the importance of détente. Still another underpinning of détente politics was his endorsement of the controversial proposition that economic development at home constituted the Soviet Union's main contribution to world revolution—in substance the Khrushchevian doctrine of peaceful competition, one of the three "peacefuls" (along with peaceful coexistence and peaceful transition to socialism) which the Chinese had violently condemned as a betrayal of Leninism during the polemical campaigns of the 1960s.

In discussing peaceful coexistence, Brezhnev signaled to the United States that it was invited to participate in the politics of détente, and he strongly underscored his regime's commitment to détente by affirming that peaceful coexistence was "not a temporary tactical method but an important principle" underlying the foreign policy of socialism. Moreover, he made the significant distinction between "aggressive circles" in the big capitalist states and those leaders who "soberly" assessed the balance of power and were "inclined to explore mutually acceptable settlements" of international issues. He made clear that Moscow would take into account "such tendencies." Inasmuch as Brezhnev's own appraisal of the international situation and of the "imperialist" adversaries' strengths was correspondingly sober, this offer to seek negotiated resolutions of basic tensions in the international system was flowing through a communications channel that had a real potential for feedback and reinforcement.

ROMANIA AND THE INDEPENDENTS

Brezhnev's blistering condemnation of the Chinese on the day after Ceausescu's plea for abstinence from attacks on absent parties posed a major decision for the Romanians, who in the event elected to remain at the conference in order to reaffirm their independent line in the Communist movement and to use their leverage to help offset the centralist pressures being generated in behalf of Moscow against Peking. In his main address, delivered on 9 June, two days after Brezhnev's, Ceausescu explained his party's decision not to walk out (as it had from the Budapest consultative meeting) but observed that Romania's doubts about the timeliness of the conference had been confirmed. Indirectly rebuking the Soviets by saying the anti-Chinese attacks did not contribute to an atmosphere conducive to resolving conflicts in the Communist movement, he justified the Romanian delegation's continued participation as an effort to do everything possible to make a contribution to Communist unity. As a neutralist on the Sino-Soviet conflict, he sought to play the role of peacemaker, recalling his party's mediatory efforts dating back to 1964 and expressing satisfaction that the feuding giants had indicated a willingness to hold talks on the border dispute. In an allusion to previous conferences' condemnation of Yugoslavia, he reminded the delegates that "in the past serious charges were also made against other parties, only to be subsequently proven to be false." Ceausescu took the occasion to offer his country's interpretation of the meaning of nationalism in the Communist movement. Where the pro-Soviet loyalists were defining nationalism as defiance of the larger interests of the Communist camp, he bracketed nationalism with hegemonism as well as with racism and chauvinism, qualities that were more suggestive of traditional Russian hegemony and pan-Slavism than of efforts by a small state to pursue its own interests. He stressed firmly that it would be "completely incorrect" to regard the assertion of the principles of national sovereignty and independence of parties as nationalist manifestations or violations of proletarian internationalism, and he explicitly noted that his discourse on the meaning of nationalism was responsive to criticism of the Romanians for neglecting their internationalist duties. Inveighing against "absolutization" of forms of revolutionary struggle or socialist construction as "universally compulsory" norms, he said this tendency was the principal source of the existing divergences in the Communist movement—a diagnosis sharply at odds with the orthodox Soviet interpretation.

In his final remarks at the conference, on 16 June, Ceausescu credited Romania with a role in shaping the main document, thus justifying the decision not to walk out. Announcing his delegation's decision to sign the document, he stressed that "improvements" in the draft had been made because "some account was taken of draft amendments presented by a series of parties, the Romanian party included." The other most important independent-minded party at the conference, the Italian CP, took another approach suited to its own situation and with an eye to popular approbation at home. The Italian delegate, Enrico Berlinguer, announced that his party would subscribe only to the main document's third section, dealing with the anti-imperialist struggle, and that its objections to the other sections ran to the "very structure" of the document. The Italians thus dramatized their resistance to any assumption of ideological authority by Moscow, particularly as was being advanced by the pro-Soviet loyalists, and gave concrete expression to the polycentrist trend most notably associated with the name of the Italian Party's late leader, Palmiro Togliatti. This was also made clear in Berlinguer's explanatory remarks, as in his observation that the document's section on the socialist countries was unsatisfactory because it outlined a uniform conception of socialism that conflicted with the type of socialist society for which his party was calling on the Italian people to struggle.

The Italian Party's autonomist propensities also shaped Berlinguer's treatment of the China question. He noted that his party had itself criticized the Chinese for their hostility toward Moscow and their pretensions to ideological authority, but he echoed Ceausescu in welcoming the fact that the conference document did not excommunicate any party, and that inter-party relations were not to be affected by whether or not any given party participated in the conference or subscribed to the main document. Asserting a proposition central to the Italian CP's conception of international Communist relations, Berlinguer deplored attempts to explain differences in the Communist movement in terms of deviation from doctrinal purity, "and no one knows who is the trustee of such purity." Such attempts, he explained, served not only to exacerbate differences but also to obscure understanding of the objective reasons and interests at the root of divergent views. Thus, in his party's polycentrist spirit, Berlinguer forcefully denied that there was a center of authority or a corpus of orthodoxy to which the various parties must submit their policies and ideas for validation. And, by the same token, he diagnosed one of the causes of the severity of the Sino-Soviet conflict as arising from efforts to ascribe divergent interests to ideological deviations.

Dissent from Moscow's line was registered at the conference not only from the reformist elements in the Communist movement, particularly as expressed on the Czechoslovak question and its implications, but also from more radical elements flanking Moscow on the "left" wing. To be sure, the absence of the radical Asian wing of the Communist movement insulated Moscow from much of the heat applied at the 1957 and 1960 conferences by the Chinese and their militant allies, but the détente policies under fire at that time were again subjected to challenge in 1969 from still another quarter of the Communist movement, that taking its inspiration from the Cuban revolutionary experience. The Cuban Party itself was present only in an observer status, represented by a second-level leader, Carlos Rafael Rodriguez. Nonetheless, its presence even as an observer reflected the upgrade in Soviet-

Cuban relations evident since the invasion of Czechoslovakia and marked another step in the absorption of Cuba into Moscow's "socialist community." Defensiveness about the decision to attend was reflected in Rodriguez's remarks at the conference, in which he noted that the negative factors in the Communist movement arguing against attendance had not been removed but that pressure from Moscow and its allies had prompted the decision to take part, lest the Cuban Party's absence might provide a propaganda weapon for imperialism. This logic comported with Castro's tortuously qualified endorsement of the intervention in Czechoslovakia and registered Havana's growing concern over being left isolated and exposed outside Moscow's protective mantle. These considerations were also reflected in Rodriguez's assessment of the situation in the Communist movement and of international affairs in general. He recorded Cuban dissent on substantial parts of the main document, but at the same time he tendered praise for the aid given by the Soviets for anti-imperialist struggles and he gave an oblique but significant show of support for Moscow against the Chinese by vowing readiness to back the Soviet Union in "any decisive confrontation," whether "concerning actions by the Soviet Union in the face of the danger of the tearing off of members from the socialist system by imperialist maneuvers [that is, Czechoslovakia] or concerning a provocation or aggression against the Soviet Union, come from where it may [e.g. China]." The latter formulation came close to Moscow's depiction of the dispute with China as part of a two-front conflict in which the Soviets were defending the interests of the Communist world. Rodriguez summarized Cuban-Soviet relations by acknowledging that there had been "inevitable discrepancies which on some occasions were very acute," while conferring on Moscow the role of "a fundamental bulwark" in the anti-imperialist struggle.

THE MAIN DOCUMENT

The CPSU's Ponomarev, reporting to the conference on its penultimate day as chairman of the editorial commission that produced the conference documents, expressed satisfaction that "a high degree of agreement" had been reached in the work of preparing the documents.[31] As to be expected, a document expressing support for the Vietnamese struggle reflected "complete unity" on an issue providing a convenient common denominator for "anti-imperialists." Ponomarev also reported that the editorial commission had been able with similar unanimity to produce a draft "Appeal in Defense of Peace," another easy achievement in view of the absence of the radical Asian parties distrustful of détente politics. As for the difficult task of producing a main document that could embrace a wide range of ideological and political orientations, Ponomarev indicated that the extensive preparations and the efforts to accommodate the various parties' views at the preparatory commission's May meeting had yielded a draft which remained largely intact through the conference deliberations. It thus seemed clear that the Romanians had seen the value of a trade-off at the conference: they remained at the conference in the face of the anti-Chinese bandwagon because they knew that a main document would emerge in which autonomist forces balanced the centralist ones. The concessions made by Moscow to achieve that balance were the price paid in order to have its cherished conference, an essentially anti-Chinese project that had been imperiled by the fallout from Czechoslovakia but had now, in the midst of the border crisis, become even more desired.

The degree of diversity Moscow faced can be measured by the varying responses to the main document by the participants. The Cuban and Swedish delegations, present only as observers, were not faced with the decision whether or not to sign the document, and the independent-minded British and Norwegian delegations deferred action; the Italian CP was joined by its pocket client party, the San Marino CP, and by the outspoken Australian Party in signing only the third, specifically anti-imperialist, section of the four-part document; the militant Dominican Party, which had taken a *fidelista* line at the conference, refused to sign any part of the document; and a dozen or so, including the key Romanian delegation, registered reservations of varying degrees of seriousness while affixing their signatures to the document. The Czechoslovak issue underlay the objections of several of these parties, as in the Australian delegation's complaint that the document disregarded important phenomena in relations among Communist countries that "negatively influence our entire movement," as well as the Romanian argument that the document overrated the capability of imperialism for creating divergences while underrating the capability of the socialist countries to repel attempts to disrupt unity (a way of denying that tighter bonds with Moscow were required). The Romanians also found the treatment of the Middle East question one-sided against Israel, whereas the Sudanese and Moroccan delegations, on the contrary, felt that the document did not sufficiently stress the Palestinians' rights. And one delegation, the Costa Rican, was more papal than the Soviet papacy in wondering "why we must keep silent [in the document] about the policy of the present leaders" of China. The trade-off mentioned above was the reason why the document did not address the China question directly; it was the reason, moreover, for the document's explicit assertion, incorporating the Romanian demand, that the world socialist system comprised 14 states, thus, by strict implication, including China and Albania. The document recorded a claim that the conference marked "an important stage in the cohesion of the international communist movement," but this was balanced by the very next sentence saying that "the absence of certain communist parties should not hinder fraternal ties and cooperation between all communist parties without exception."[32]

Facilitated by the absence of the radical Asian wing, Moscow was able to achieve a firm consensus on issues of war and peace, issues that had lain at the heart of the tensions that were only precariously contained at the 1960 Conference. In his authoritative commentary on that conference, Khrushchev had given a graphic portrayal of the destructive effects of a world war in the nuclear age, thus justifying Moscow's détente policies in the face of Chinese challenge. Such a portrayal was embodied in the 1969 statement, which said a nuclear war would "annihilate hundreds of millions of people and turn entire countries into deserts." Similarly, Khrushchev's commentary had been more forthcoming than the 1960 statement regarding the likelihood that Western leaders would be responsive to the policy of coexistence. The Soviet view found expression in the 1969 statement, which depicted deepening contradictions within Western "ruling circles" between belligerent extremists and those who "tend to take a more realistic approach" to international issues. The commitment to détente, which Brezhnev in his address had insisted was a strategic principle rather than a mere tactical move, was recorded in the 1969 statement as follows: "The main link of united action of anti-imperialist forces remains the struggle for world peace and against the menace of thermonuclear war and mass extermination, which continues to hang over mankind." The China problem, now dramatized in the border crisis, had become an incentive to détente politics with the West.

The policy of peaceful coexistence having provided a broad rallying ground for consensus, the most contentious issues for the conference participants revolved around the balance between centralist and autonomist tendencies in that broad sweep of the Communist movement still in communion with Moscow. That no leading center of the Communist movement would be acknowledged was a foregone conclusion, having been conceded by Moscow at the initial February 1968 consultative meeting. The most crucial ideological issue, which Moscow had diagnosed as the root of both the China and Czechoslovak problems, was that of nationalism in the Communist movement. Significantly, the 1969 statement offered a balanced proposition on the relation between international and national responsibilities that contained no mention of nationalism as an ideological deviation:

> Each communist party is responsible for its activity to its own working class and people and, at the same time, to the international working class. Each communist party's national and international responsibilities are indivisible. Marxist-Leninists are both patriots and internationalists. They reject both national narrow-mindedness and the negation or underestimation of national interests, and the striving for hegemony.

The conference delegations could not have been unmindful of Moscow's ideological campaign against nationalism or of the formulation that had appeared in the 1960 statement's treatment of the balance between national and international considerations. The 1960 statement, in reiterating the 1957 declaration's line that there must be neither undue emphasis on national characteristics nor "mechanical copying" of other countries' policies, had added a condemnation of "nationalism and narrow-mindedness." Not only was the reference to nationalism conspicuously absent from the1969 statement, but the addition of the rejection of hegemony seemed to come straight out of Ceausescu's discourse at the conference on the proper meaning of internationalism.

The significant influence of the Romanian and other independent-minded parties was also reflected in the document's treatment of ideological deviations. The 1957 declaration, composed at a time of strong centralist pressures, had identified revisionism (the reformist deviation) as the main danger to the Communist movement, and the 1960 Conference had confirmed that thesis while balancing it with a strong warning against the threat of dogmatism and sectarianism (the leftist deviation). In his address at the Budapest consultative meeting, Suslov had decried "dangerous nationalistic tendencies" appearing in the Communist movement, and his speech marking the sesquicentennial anniversary of Marx's birthday in May 1968 had put nationalism first among deviations in calling for struggle against nationalist, dogmatic, and revisionist distortions of Marxist-Leninist doctrine.[33] But the 1969 Conference statement, avoiding any mention of nationalism, called upon Communists to "uphold their principles and work for the triumph of Marxism-Leninism and, in accordance with the concrete situation, fight against right and left opportunist distortions of theory and policy, against revisionism, dogmatism, and left sectarian adventurism."

That the 1969 Conference document should have had such a Romanian imprint had considerable implications for the international system. A small country, a member of Moscow's alliance system and an immediate neighbor of the Soviet Union, had been able to exercise substantial leverage in the face of Moscow's steam-rolling influence and in the face of a recent brutal show of Soviet willingness to club a wayward ally back into submission. The effects of the Sino-Soviet schism on the constraints of bipolarity in the international system had conferred on the Romanians a leverage vastly out of proportion to their country's inherent geopolitical weight. Peking, acting from its perception of the systemic implications of Czechoslovakia, had now developed an assertive East Europe policy in an effort to enhance that leverage. For Moscow, the China problem had now appeared with still greater force as a complication of bipolar politics, and the effects of that complication had been evident in the 1969 Conference results. In short, the China problem was Romania's opportunity. It was soon to be seen that this was an opportunity available to others, and the seizure of such an opportunity by the United States dramatically affected the shape of international politics.

THE SOVIET APPRAISAL

Moscow's hopes from the conference, and the uses to which it intended to put the conference results, can be measured by subsequent Soviet interpretations of the meaning of the 1969 Conference. A CPSU Central Committee Plenum, convening on 26 June to hear a report by Brezhnev on the conference results, adopted a resolution hailing the conference as "a great success" marking "an important stage on the road of strengthening the solidarity of the international communist movement on the principles of Marxism-Leninism and proletarian internationalism."[34] That appraisal was an elaboration of the formula in the conference statement calling the meeting an important stage in the cohesion of the Communist movement—without, of course, including the statement's Romanian addendum that the absence of other parties should not impair relations among all CPs without exception. The cause of solidarity, the CPSU resolution said, should be served chiefly by "unswerving implementation of the principles of socialist internationalism, correct combination of the socialist states' national and international tasks, and the development of fraternal mutual aid and mutual support. The conference declared with great force that defense of socialism is the internationalist duty of communists." Thus, the resolution sought to infuse into the conference results those prescriptions of pro-Soviet "internationalism" that the Romanians and their ilk had so substantially diluted in the conference statement.

Most strikingly, the plenum resolution sought to read into the conference results the ideological diagnosis of the ills of the Communist body politic that had been conspicuously and significantly absent from the statement. According to the resolution, the CPSU Plenum attached "great importance" to the "conference's conclusion" that "a consistent struggle for the *purity* of Marxism-Leninism and against revisionism, dogmatism, and *nationalism* is a necessary condition for strengthening the ranks of the communist parties" (emphasis added). But the autonomist parties had successfully resisted any move to invest in the conference the authority to speak in the name of doctrinal purity, and the CPSU resolution's formulation of the types of deviation confronting the Communist movement was a blatant rewriting of the conference statement's specification of "revisionism, dogmatism, and left sectarian adventurism" as the deviations to be combated. The CPSU's tampering with the conference main document—surely the certified repository of the conference's conclusions—provided a strong signal of Moscow's determination to attempt to check independent behavior in the Communist movement.

As to be expected, the CPSU Plenum also signaled Moscow's determination to press the anti-Chinese campaign with full fervor. The plenum noted that "the exchange of opinions at the conference convincingly showed that the foreign policy line of the present leadership of the CCP and its splitting policy are encountering a resolute rebuff from the overwhelming majority of the fraternal parties." That "exchange of opinions" was what Moscow had orchestrated in the chorus of anti-Chinese addresses voiced by two-thirds of the delegations. According to the CPSU resolution, Moscow would "wage an uncompromising struggle against the anti-Leninist ideological lines of the present leaders of China, against their schismatic course and great-power foreign policy." More broadly, the plenum pledged the CPSU to pursue the line of unifying the international Communist movement "on the principled basis of Marxism-Leninism and proletarian internationalism" and to "wage a struggle against bourgeois ideology, for the purity of Marxist-Leninist doctrine, and against right and 'left' revisionism and nationalism." It was that variety of Communist unity that the Romanian, Italian, and other such parties were determined to forestall.

An important message to the Communist movement in the wake of the conference was conveyed in the form of an article by Brezhnev appearing in both *Kommunist* and the August issue of *Problems of Peace and Socialism*, the organ of the international movement.[35] The article put the conference in the perspective of a schismatic decade, taking note of the Chinese challenge since the beginning of the decade and the intensification of "leftwing and rightwing opportunism" as posing an "urgent task of erecting a barrier on the path of centrifugal trends" in the Communist movement. Brezhnev made clear the central place occupied by the China question in coping with these centrifugal trends. In his interpretation of the conference's treatment of this question, he said it "clearly showed the danger of the splitting activity of the present CCP leadership for the international communist movement, and the enormous damage it does to the anti-imperialist struggle." After noting that various delegations had condemned the "attempts to replace Marxism-Leninism with Maoism," he claimed that it could "definitely be said that the conference resolutely condemned this ideology and policy"—a claim belied by the resounding silence of the conference main document on the China question. Significantly, however, Brezhnev at the same time offered the assurance that the Soviet Union would not allow itself to be provoked into rash action in dealing with the Chinese challenge.[36]

A note of restraint was also evident in Brezhnev's discussion of the ideological issues raised by Moscow's efforts to cope with centrifugal trends. He frankly acknowledged the limits of Moscow's control over the conference delegations, noting that several assented only to the section of the conference statement dealing with anti-imperialist united action, and that some parties signed the document but expressed reservations. Projecting an ecumenical spirit, he insisted that the "main thing" was that the conference yielded a unity of stand of "nearly all participants on a wide range of urgent and basic problems." To be sure, he raised the controversial issue of nationalism in saying that the danger of revisionist deviation was aggravated by the fact that in a number of cases it also fused with bourgeois nationalism; but, more significantly, he called for a differentiated approach to the ideological struggle for the purity of Marxism-Leninism. He explained this differentiation as being one between struggle against the class enemy and struggle directed at allies who were "temporarily deluded." In the latter situation, "comradely, friendly polemics" were useful, and at times "a certain restraint." In short, Moscow was concerned to keep open its lines of communication with a broad spectrum of the Communist movement in an effort—complementary with détente politics toward the West—to isolate and contain the aggressive challenge of the Chinese.

NINE

The Summer of Sixty-Nine

The Chinese Party Congress in April 1969 and the Moscow International Party Conference the following June helped define the political context in which the border crisis was evolving. Having agreed to the convening of the border river navigation commission, Peking next made a response to Moscow's proposal to open discussions on the border dispute as such. On 24 May, Peking issued a Government Statement on the subject,[1] thus matching the authoritative level of the Soviet statement eight weeks earlier and laying out the fundamentals of the Chinese position less than two weeks before the opening of the Moscow Conference. It became evident during this period that both sides were playing to the international Communist gallery, including those parties boycotting the Moscow Conference as well as those participating. In its statement, Peking assented to opening border talks, at a time and place to be decided through diplomatic channels, but the negotiating position set out made abundantly clear the fundamentally divergent approaches to the issue—even to definition of the issue—being taken by the antagonists. So divergent were their approaches that it would take several more months, in a crisis situation, before they could come together at the negotiating table to discuss the territorial question.

The Chinese statement insisted that there was an outstanding territorial question arising not only from the "unequal" treaties imposed on a weak China by tsarist Russia but also from occupation of land lying on the Chinese side of the border as stipulated by those treaties. In the latter category the statement included 20,000 square kilometers in the Pamir area of Central Asia as well as more than 600 of 700 islands on the Chinese side of the central channel of border rivers. It said the Soviet side in the 1964 talks had admitted that the central channel determined ownership of islands, a concession the Soviets would be loath to renew after the fighting for possession of one of those islands. In charging the Soviets with violations of the border since 1960, the statement cited the notorious incident in the spring of 1962 in which the Soviets allegedly "incited and coerced" more than 60,000 Chinese citizens (of non-Han minority nationality) into going into the Soviet Union.[2] Bringing precision to their charges, the Chinese claimed that the Soviets had provoked as many as 4,189 incidents in the period from 15 October 1964 to 15 March 1969, a total two and a half times the number of incidents in 1960-1964. This was a cleverly turned barb in response to Moscow's reference to 15 October 1964 as the date on which the border talks were to have resumed; like Moscow, the Chinese neglected to mention that this was one day after Brezhnev formally replaced Khrushchev as CPSU First Secretary.

In the operative section assenting to a reopening of discussions on the border question, the statement pointed to a crucial divergence in the two sides' definition of what in fact was at issue. According to the statement, Moscow's professed willingness to resume "consultations" reflected an attempt to deny the existence of a border question, "which actually amounts to saying there is nothing to discuss at all." The Chinese were thus making explicit Moscow's resistance to anything that would appear to open the border to renegotiation, as distinguished from an effort to delineate more precisely sections of the border that were unclear or had undergone natural changes. As for Moscow's appeal for measures to be taken to normalize the tense border situation, the Chinese statement disclosed something about the situation around Chenpao since the two fire-fights had erupted. The statement noted that the Soviets had continued to fire on the island and that "to this day the firing had not ceased." The Chinese also claimed that a Soviet border representative on 3 April had read a prepared statement saying the firing would not cease until the Chinese agreed to hold talks on the matter and unless the Chinese withdrew from the island. This confirmed other indications that, after the March clashes, the Soviets had not attempted to occupy the island and thus were avoiding a repetition of the fire-fights. This was also another sign that Moscow, regarding the border dispute as a political and diplomatic liability, was seeking to defuse tensions. However, the statement at the same time indicated that troubles were brewing elsewhere, charging that the Soviets were "carrying out provocations in other sectors" of the border. In addition, the statement complained that the Soviet propaganda machine had been set into motion to foment chauvinistic feelings and that the Soviets had "brandished nuclear weapons at China." As has been noted, it was especially in Soviet broadcasts to the Chinese that the specter of a large-scale conflagration involving Soviet strategic power had been evoked to dissuade the Chinese from border adventures. All of this behavior by the Soviets, the statement warned, made it "highly doubtful as to how much sincerity" Moscow had for negotiations.

The Chinese statement set out tough terms for the negotiations, terms that clashed in fundamental respects with the Soviet position. Peking insisted that "it must be confirmed" that all the treaties relating to the border were unequal, a demand for redress of historical grievances that Moscow was utterly loath to accommodate, lest a dangerous precedent be set for reopening territorial questions with all sorts of countries. Nonetheless, the Chinese statement added, Peking remained prepared to take the inherited treaties as the basis for determining the entire border alignment, thus indicating that the historical grievances were not being made the basis for an irredentist claim. But there would need to be negotiations on an overall settlement and the conclusion of "a new equal treaty" to replace the old unequal ones, in contrast to Moscow's offer merely to hold consultations on specific stretches of the border. Though Peking insisted that any side occupying territory in violation of the treaties "must, in principle, return it wholly and unconditionally to the other side" (and "this brooks no ambiguity"), it also offered to make "necessary adjustments at individual places" on the basis of mutual understanding and accommodation. This would appear to have provided a necessary point of convergence between the two sides' positions, offering an area for real bargaining and negotiation. However, in defining the status quo from which such bargaining could proceed, the Chinese again had recourse to an ambiguity in their appeal to the line of actual control. The statement proposed that each side should maintain the status quo "and not push forward by any means the line of actual control on the border, *and* that in sectors where a river forms the boundary, the frontier guards of its side shall not cross the central line of the main channel" (emphasis added). The latter element in this formula introduced a juridical element in addition to the notion of actual control, thereby begging part of the question. Chinese patrolling of disputed islands like Chenpao lying on the Chinese side of the main channel represented an effort to resolve this ambiguity by making the line of actual control conform with the legal principle of the thalweg. This consideration helps explain the tenacity with which the Chinese had sought to maintain their position on Chenpao.

Having expressed Peking's willingness to hold border talks, the statement also warned the Soviets not to take this readiness for a peaceful settlement as a sign of weakness, as a basis for thinking that the Chinese could be cowed by "nuclear blackmail" (again that sensitivity to Moscow's use of its strategic superiority to threaten the Chinese into line). But Peking also used the occasion to signal even more strongly its aversion to becoming embroiled in a dangerous conflagration. In addition to repeating Mao's directive against attacking unless attacked, the Chinese said that, "as far as our own desire is concerned, we don't want to fight even for a single day. But if circumstances force us to fight, we can fight to the finish." The considerations underlying these signs of caution were reflected in an NCNA report on 2 June accusing the Soviets of making use of nuclear threats as part of a dual approach of "political deception and military adventurism." NCNA took note of Soviet assertions that nuclear arms formed the main weaponry and that missile units stationed in the Transbaykal region and along the Sino-Mongolian border had been put in battle array to give a "crushing nuclear rebuff" to the Chinese.[3] Clearly there was a dangerous and delicate process of signaling taking place in a destabilized setting.

In publishing both the Chinese statement and the Soviet one of 29 March to which it was a reply, Peking also disclosed its response to the 11 April Soviet note urging the Chinese (at that time assembled in the Party Congress) to arrive in Moscow four days later for "consultations." The Maoist touch was evident in Peking's response dated 14 April: "We will give you a reply; please calm down a little and do not get excited." Now, in late May, the reply had been presented, and so the ball was back in the Soviet court. And the next Soviet move on the negotiations front would take place under the concentrated gaze of 75 foreign Communist delegations in Moscow for the International Party Conference.

The formal Soviet reply to the Chinese Government Statement came on 13 June, but six days earlier Brezhnev made the first public acknowledgment of the 24 May statement in the course of his extensive anti-Chinese discourse before the international conference.[4] Moscow had previously agreed not to air the China question at the conference, but Brezhnev's polemical drumbeat set the pace for an anti-Chinese march that turned the conference into a pro-Moscow rally in the Sino-Soviet conflict. In his major address on the third day of the conference, which opened on 5 June, Brezhnev justified the controversial move to raise the China question by citing the "grave negative influence" of Chinese policies, especially as registered at the April CCP Congress, on the international situation. As for the border question, he cautiously welcomed Peking's indication that it would hold talks and wished to avoid border clashes, and he revealed that the Soviets were preparing their formal reply to the Chinese statement. But he took the occasion to complain that the latter could "hardly be described as constructive" and was full of "historical falsifications." Notably, Brezhnev stressed that the Chinese had renewed "groundless territorial claims" which the Soviets "categorically" rejected. He was, thus, treading carefully in order to leave open the way for talks without conceding the existence of a territorial question as such. In another context looking toward a long-term strategy for coping with the China problem, Brezhnev briefly and without elaboration floated a proposal for an Asian collective security system. He introduced this significant if inchoate signal in connection with the longstanding proposal for a European security conference that would ratify and solidify the postwar settlement in Europe, particularly as regards the German question. In now suggesting the possibility of a parallel project for Asia, Moscow was hoping to launch a campaign for consolidating the postwar Asian territorial arrangements as part of its effort to cope with the China question.

The lengthy 13 June Soviet Government Statement[5] offered a point-by-point rebuttal of the Chinese negotiating position, while at the same time welcoming a move toward defining an area of convergence in which a process of accommodation might be pursued. The statement complained that the Chinese were not creating a favorable atmosphere for talks by seeking to open discussion of a new border treaty rather than continuing the 1964 "consultations" on a more precise boundary demarcation. Rejecting Peking's arguments that the Soviet regime in Lenin's time had renounced the border treaties, the statement flatly declared that "there is no territorial question" outstanding and that the borders were "inviolable today as they were inviolable yesterday. Any attempt to cross the Soviet border will call forth a crushing rebuff." In this connection, the statement charged that the Chinese had continued their provocations along the border.

As for the disputed islands, the Soviet statement repeated the claim that 1861 protocols to the Treaty of Peking drew the boundary line along the Chinese bank of the Ussuri in the Chenpao area. Trying to lend stronger legal force to their argument, the Soviets countered Peking's invocation of the thalweg principle by pointing out that in international law there is no rule automatically establishing the central channel of navigable border rivers as the boundary but that states determine the boundary as appropriate. It cited as examples the 1860 Treaty of Peking itself and an 1858 treaty between Costa Rica and Nicaragua in which the boundaries were drawn along the river bank instead of along the mainstream. In introducing this argument, the Soviets did strengthen their case on an issue that was pivotal to Peking's position. Depending on the validity of the supplementary protocols being cited, this argument has some merit. In international law, the boundary line as a rule runs along the main channel of navigable rivers, but in exceptional cases arising under treaty or long-established peaceable occupation the boundary can lie along one bank of the river, giving sovereignty over the whole river to the other riparian state.[6] But Moscow's counterargument left the issue inconclusive, inasmuch as the Chinese had claimed that the 1861 protocols and map indicated only that the Amur and Ussuri formed the border but did not specify that the boundary ran along the bank.

More significant than these forensic points, however, were indications of a bargaining approach toward issues that were more likely to be subject to real negotiation. The Chinese having claimed that the Soviet delegation at the 1964 talks had acknowledged the main channel as determining the boundary, the Soviet reply now disclosed that in those talks the Soviet side had stated its readiness "to meet halfway" the wishes of the Chinese in order to reach agreement on the boundary along the Amur and Ussuri rivers, this to be on the basis of mutual concessions and on condition that the Chinese in turn showed readiness to recognize the interests of the Soviet population along separate sections of the border. This indicated that the Soviets had been prepared to accede to Chinese demands with respect to at least some of the riverine islands, provided that Soviet rights and interests be protected on those islands of particular importance to the Soviet Union. But such an accommodation was not in the cards dealt in the 1964 talks because, in the Soviet account, the Chinese "complicated consultations by making groundless territorial and other claims that questioned both the line of the existing border and all treaties determining the Soviet-Chinese border." In other words, the Chinese for their own purposes were insisting on opening the border to renegotiation rather than engaging in concrete consultations looking toward an accommodation in accord with local conditions and interests.

In the operative section of the statement, Moscow renewed its offer to resume "consultations" in order "to discuss the question of specifying the boundary line on individual stretches, proceeding from the border treaties in force," and without preconditions. Having thus rejected the question of unequal treaties and of formal renegotiation of the border (even with the inherited treaties as the basis), the statement proceeded to outline the steps the Soviets envisioned as a way toward an accommodation of the border conflict. It proposed that the two sides first record agreement on undisputed sections of the border, and then "reach understanding on the boundary on separate disputed sections through mutual consultations on the basis of treaty documents." Further, proceeding "from the treaties in force," the two sides should observe the "principle of mutual concessions and the economic interests of the local population when delimiting the boundary line on sections which have undergone natural changes." Finally, the agreements reached should be recorded "by the signing of appropriate documents" by the two sides. Having thus artfully sought to focus the proposed talks on concrete bargaining over specific parcels of land (on which the Soviets could afford to be generous) rather than on a territorial question as such (which the Soviets dreaded as a dangerous precedent), Moscow welcomed the approach taken in the Chinese statement that would seek necessary adjustments in individual sections of the border on the basis of existing treaties. Moreover, the statement endorsed the Chinese suggestion that, in the meantime, the two sides should avoid conflicts along the border and should not resort to arms.

In an attempt to put pressure on Peking to move toward the bargaining table in the near term, the Soviet Statement proposed that the 1964 "consultations" be resumed in Moscow within the next two to three months, and it added that the Soviet Government expected the Chinese to inform it "shortly" whether these proposals on time and place for the talks were acceptable. As still another element of this package of concrete arrangements, the statement named P.I. Zyryanov, who had headed the Soviet delegation in 1964, to be the delegation leader with the rank of Deputy Minister. As chief of the Soviet border guards, he had recently written an article in *Pravda* marking Border Guards Day (28 May) in which he stressed Moscow's determination to resist a renegotiation of the border. The article emphasized that the Sino-Soviet border was established "long ago, lies along natural boundaries, and was legally formulated in the appropriate treaties" which "retain their force" today.[7] Zyryanov's position as border guards chief (as distinguished from a political or diplomatic position, such as a foreign ministry official) comported with Moscow's approach that specific adjustments along a legally determined border were all that would be on the agenda for consideration. The Soviets had thus carefully packaged their proposals to conform with the basic premises on which they were prepared to discuss the border with the Chinese.

The Chinese Government Statement of 24 May had referred in passing to troubles in "other sectors" of the Sino-Soviet border, thus suggesting that the situation had become enflamed in the volatile Central Asia area. This ominous possibility was confirmed when Peking on 6 June, the second day of the Chinese-boycotted International Party Conference in Moscow, issued a Foreign Ministry note[8] containing the sweeping charge that the Soviets had extended their border provocations from the Ussuri to the Amur, from the riverine to land borders, and from the eastern to the western sector. The note cited several incidents in April and thereafter in which the Soviets had allegedly violated the status quo and interfered with Chinese patrols in Sinkiang. Incidents were also mentioned as taking place on islands in the Amur lying on the Chinese side of the central channel, including one in which the Soviets fired at a Chinese patrol and killed one border guard. In addition, the Soviets were accused of having penetrated Chinese airspace 57 times from 29 March to 31 May, some of these penetrations extending to a depth of 60 kilometers. This was quite a panoply of charges to hurl at Moscow while representatives of 75 Communist parties were assembled there for what turned out to be a clamorously anti-Chinese forum.

The Chinese note was calculated to show—to the international Communist audience in particular—that tensions had persisted and extended across the vast extent of the Sino-Soviet border, though nothing of the seriousness of the March clashes was indicated. Something approaching those clashes, however, was signaled by Peking's prompt release on 11 June of a protest note concerning an incident along the Sinkiang border on the previous evening.[9] According to the note, fighting had developed between the opposing border units after the Soviets had opened fire and killed a herdswoman; the note said the situation was still "developing," an observation that the Chinese had also made regarding the major 15 March clash and which indicated the serious scope of the incident. Moscow had remained publicly silent about the omnibus Chinese charges in the 6 June note, but a Soviet Foreign Ministry note on 11 June [10] sought to put the record straight on the latest incident by explaining that the Chinese had used a herdsman and a flock of sheep as cover for a group of soldiers who suddenly opened fire in reply to Soviet border guards' demands that the "trespasser" leave Soviet territory. Taken together, the two accounts suggested that the Chinese were following their practice of contesting Soviet jurisdiction over disputed territory and using force to counter attempts to expel them. According to the Soviets, an earlier but unpublished Soviet note on 4 May had conveyed a "strict warning" against Chinese provocations along this sector of the border. The 6 June Chinese note had also mentioned the 4 May note and two others that Moscow had not released publicly. According to the Chinese, there had been an incident on 2 May in the same area as the 10 June clash, also involving herdsmen and flocks. These subsequent accounts thus confirmed the significance of articles in both *Pravda* and *Literaturnaya Gazeta* on 7 May which indirectly signaled the rising tensions in the Central Asian borderland. Using the devise of interviews with refugees from Sinkiang to recount tales of oppression of the Turkic minority peoples, one article blatantly fanned separatist sentiments among those peoples by such means as citing a desire to rename Sinkiang "Uighurstan";[11] the other article, stoking disaffection in Sinkiang, contrasted a flourishing life in Soviet Central Asia with backward conditions on the PRC side.[12] Together with the diplomatic warnings, all of this amounted to a minatory message to Peking that Soviet capabilities for troublemaking in Sinkiang would make that region a dangerous arena for border confrontations.

The 10 June clash having confirmed Peking's reference four days earlier to border tensions in Central Asia, an incident on 8 July did the same for the reference to trouble on the Amur. As in the case of the 10 June incident, Peking was the first to issue a protest, claiming that two Soviet vessels landed border guards on Pacha (Goldinskiy) island who opened fire on "Chinese militiamen and inhabitants engaged in production" there.[13] Peking claimed title to the island on the ground that it lies on the Chinese side of the central channel; the incident thus provided another example of Chinese efforts to establish actual control over disputed territory. According to a Soviet note released a few hours later,[14] a group of armed Chinese, hidden in ambush "on the Soviet part" of the island, had fired upon Soviet transport workers who had come to repair navigation markers in accordance with usual practice. The note claimed, probably correctly, that the Chinese knew well that the navigation markers had been established and always maintained by the Soviets, and it warned that the Soviets would be compelled to take "additional measures" of protection against Chinese actions endangering the security of Soviet citizens (one was killed and three wounded in the incident, the note said). The incident thus seemed to portend further troubles, with the Soviets intent on countering challenges to their jurisdiction over border river islands.

Moscow's reference to the Soviet "part" of Pacha was a curious one and may have reflected uncertainty over how to handle the question of title to the disputed islands. If Moscow were to accept the thalweg principle, then it would logically follow that an island belonged fully to one or the other side. However, the Soviets were resisting that principle and were offering to bargain with the Chinese on a more ad hoc basis, one that would avoid the broader territorial issue in favor of specific adjustments of the border demarcation. On 9 July, TASS carried an account of the Pacha incident that said the island was a Soviet one on which Chinese peasants were permitted to engage in farming by agreement of the two sides—a practice comporting with Moscow's position on the border question. But as an NCNA report five days later mockingly pointed out, TASS had issued another dispatch one hour later amending the story to say the Chinese peasants were permitted to farm on the Soviet "part" of the island, an emendation that would put the follow-up story in harmony with the official protest note. NCNA sarcastically noted that even Western news agencies had seen through the Soviet story and raised the question whether Moscow intended to claim ownership of all or only part of the island.[15]

If even the Western press could notice the discrepancy, then why did it take the Chinese five days to point it out? The timing of the Chinese reaction almost surely related to another pair of TASS reports indicating, this time, uncertainty on the part of the Chinese regarding how to treat border-river questions. Successive TASS dispatches, the first transmitted shortly after midnight and the second in the afternoon of 13 July,[16] disclosed that the Chinese had refused, on the 12th, to continue this year's session of the joint commission on border river navigation (which had opened on 18 June) and then the next morning had agreed to resume it on the 14th. The first dispatch, reporting the walkout decision, said that at the session the Soviets had submitted a plan for traffic maintenance work in 1969 and a proposal for discussing a new text of the navigation rules for ships using the border rivers. However, according to TASS, it became clear from the beginning that the Chinese delegation sought to impose on the commission questions concerning the location of the boundary, an effort demonstrated even during the discussion of the wording of the agenda; the Chinese had also raised the issue of unequal treaties, while the Soviet delegation sought to limit discussion to matters specifically relating to navigation. TASS repeated the charge made in the 8 July Soviet note that the Pacha incident was staged by the Chinese in order to undermine the work of the commission. The Pacha incident involving Soviet navigation personnel, taken together with the abortive walkout from the joint commission's session, suggested that there were important elements in Peking intent on keeping the border issue alive and resisting Soviet efforts to use the navigation talks as a precedent for arriving at accommodations based on limited, specific agreements. That the Chinese were actively contesting Soviet jurisdiction along the border was reflected in a speech at an 11 July session of the Supreme Soviet in Moscow by a representative from Khabarovsk Kray (the joint navigation commission was meeting in Khabarovsk), who noted that no results had been achieved after 18 meetings of the commission since 18 June. He charged that the Chinese had already been guilty of more than 130 border violations that year and that Chinese fishermen, "clearly acting on orders" from the authorities, had continued attempting to fish in Soviet waters; 113 Chinese fishermen had been "turned out of Soviet waters" in the Amur and the Ussuri during the preceding two months.[17]

Despite the conflicting signals and the fundamental divergence in the two sides' conception of the joint commission's scope, the session was resumed and was able to achieve at least some agreement, a result signifying some give in the most intransigent elements, particularly on the Chinese side. TASS promptly reported on the closing day, 8 August, that, after discussions of "practical questions" concerning shipping conditions, the co-chairmen signed a protocol recording agreement on "certain measures to improve the shipping situation." The two sides also agreed to hold the next session in China in 1970, thus appearing to regularize an activity that had been threatened by the Chinese insistence on injecting territorial questions.[18] Understandably, in view of the conflicting signals from the abortive walkout, the Chinese were less eager to publicize the positive results, not reporting the conclusion of the commission's session until coyly announcing the return to Harbin of the Chinese delegation on 11 August.[19] Implying that the Chinese made some concessions, NCNA said the Chinese delegation, "displaying the attitude of making the conference a success and seeking solutions to the problems, patiently conducted negotiations" and signed a protocol recording agreement on "some concrete questions." At no time did Peking confirm or deny the Soviet report of the Chinese walkout because of Soviet refusal to take up territorial issues. The Chinese decision to return to the session and to agree on some concrete measures indicated some significant loosening of the Chinese demand in June that the commission discuss "all the questions" put forward. On the other hand, the results of the session fell short of Moscow's proposal of a new text for the 1951 agreement on border-river navigation, but that would have involved far too great a sacrifice by the Chinese hard-liners.

With some measure of accommodation being achieved on the border rivers, and with the lack of an ice covering during the summer reducing the likelihood of clashes there on the scale of the March fire-fights, the area of greatest potential trouble at this time was in Central Asia. As was shown in the early 1960s, the possibilities of troublemaking among restive minority peoples gave this area an especially explosive potential, and the location in Sinkiang of Chinese nuclear and missile facilities increased that potential. It will be recalled that the visit of the Albanian military delegation in October 1968 to Sinkiang, in itself an unusual event, had occasioned a notably explicit warning of a possible Soviet attack and of the need for preparedness. After the serious incident on 10 June, the Chinese journal *Red Flag*[20] carried an article attributed to two Sinkiang locals, a peasant who was a token member of the CCP Central Committee and a herdsman, stressing the importance of unity in Sinkiang in the interest of raising the level of combat readiness there. After referring to Chenpao and the recent Sinkiang clash as well as reciting again the 1962 incident in Ili, the article conveyed the exhortatory message that "we will transform the raging flames of anger against the Soviet revisionist new tsars into a mighty force for uniting together all nationalities." In this effort to arouse a sense of identity with the Peking regime, the article took wary note of efforts by the Soviets and local elements to promote tensions between the minority nationalities and the Han Chinese, including the use of religion. Also at the time, Peking sought to turn back on Moscow the charges of oppression of minorities, as in an NCNA report on 1 August portraying Soviet mistreatment of minority nationalities in Central Asia. Even more notable, however, was the NCNA report's reference to Soviet missile units in that region, one of a panoply of charges that the Soviets had set up missile launching sites, nuclear weapon testing grounds, military bases, and strategic highways, and had concentrated large numbers of troops there.[21]

The Chinese seem to have had good reason to point with alarm to menacing Soviet behavior in Central Asia. The Chinese had fought a border war with India growing out of Peking's effort to control a distant borderland, and the problem of control over Sinkiang had long since become a matter of grave concern and a sharp irritant in Sino-Soviet relations. One of the more ominous signs of Soviet troublemaking in mid-1969 was the renewed publicity surrounding the Uighur refugee and former general in the Chinese army, Zunun Taipov, who had figured in the menacing Soviet signaling of Khrushchev's last years. In an article in the Soviet foreign affairs weekly *New Times*[22] bearing the provocative title "Maoist Outrages on Uighur Soil," he recalled with threatening overtones his participation in the 1944–1949 uprising by Uighurs and other non-Hans against Chinese authority and the establishment of the East Turkestan Republic. In a notably blatant probing of separatist sentiment in Sinkiang, the article attributed the establishment of the non-Chinese government in Chinese Turkestan to a "Uighur people's national liberation movement," a characterization serving to legitimize anti-Chinese revolts. Lest the relevant message be missed, the article updated the story with a reference to the 1962 Ili incident and a portrayal of continuing resistance in Sinkiang to the Peking regime.

Given the long history of Russian pressure eastward in Central Asia, a memory exacerbated for the Chinese by the 1962 incident, it was scarcely to be wondered that signals like the Taipov article, amplified by broadcasts from transmitters in Soviet Central Asia, were regarded by Peking with serious alarm. Having reversed their original appraisal of the 1959 Tibetan revolt by treating it now as an expression of legitimate grievances, and now also playing up separatist tendencies in Sinkiang as equally if not more legitimate, the Soviets may very well have been tempted to exploit restiveness among the minorities in Sinkiang in an effort to promote there an insurgent movement challenging Chinese control. In such a scenario, a figure like Taipov, a Uighur native of Sinkiang with a past record of participation in anti-Chinese uprisings, would be ideally cast for the role of leader under Soviet direction of a "national liberation movement" against "great-Han colonialism." The prospect tempting the Kremlin would be one of a reversion of Sinkiang to its status of a virtual satellite during the Stalinist era; or, to use another relevant example, of a conversion of Sinkiang to the satellite status of Outer Mongolia. Peking may have been willing to reconcile itself to the alienation of Outer Mongolia, but its forceful moves against India in seeking to secure control over Tibet had testified to the Chinese determination to maintain possession of Sinkiang.

It was, thus, in a highly charged atmosphere that the most explosive confrontation since March erupted along the Central Asia border on 13 August. This clash broke a month-long lull in reports on border incidents and came five days after the joint navigation commission had concluded its work with a measure of success. A Chinese Foreign Ministry note,[23] released an hour before Moscow's, claimed that several hundred Soviet troops accompanied by helicopters and tanks intruded 2 kilometers into the Tiehliekti area of Sinkiang's Yumin county and opened fire on Chinese border guards conducting "normal patrol duty." An unspecified number of Chinese border guards were killed or wounded, according to the note. As in the protest over the 10 June incident in the same county, the Chinese note said the situation was "still developing" and that the Soviets were continuing to mass large numbers of troops and tanks. That this incident had a long background was indicated in a subsequent Chinese report charging that, in recent years, the Soviets had on frequent occasions similarly massed troops and armor in this area and points north and south along the Sinkiang border.[24] The report said that, in the summer of 1967, the Soviets had used bulldozers to destroy a 7-kilometer-long patrol route in this area, and that, in June 1969, they penetrated to a depth of 3 kilometers while erecting installations and establishing boundary markers.

The Soviet Foreign Ministry's note[25] reviewed the earlier warnings delivered in the spring concerning Chinese attempts to establish a presence in territory claimed by the Soviet Union. According to the Soviet version, their border units on 12 August observed the Chinese moving military units into this area and laying communications lines; the Soviet border authorities requested a meeting with a Chinese representative, but this was rejected. On the 13th, the Soviet note recounted, "several groups" of Chinese troops violated the border 10 kilometers east of Zhalanashkol and opened fire after the Soviets gave warning signals and requested them to withdraw. Paralleling Peking's claim that the Soviets had reinforced their troops, Moscow's note said the Chinese tried to move in two more groups of 60-70 troops after the opening fire; the Soviets said they threw the Chinese back, with "several" casualties. The note promised a "firm rebuff" to any encroachment on Soviet territory. The 13 August incident had revealed an alarming potential for escalation to a level of hostilities that would be difficult to control.

In the immediate aftermath of this incident, the Chinese went to some length to portray intensified Soviet propaganda and military mobilization against China, including heavy reinforcements and construction of strategic highways and railroads along the border. There may, however, have been some uncertainty in Peking over how far to go in portraying menacing Soviet behavior—an uncertainty perhaps reflecting some counsel of caution. Thus, one NCNA report included a passage detailing the construction of roads and railways and shipments of troops and equipment but, an hour and a half after the report was transmitted, an instruction to delete this passage was issued by NCNA. But Peking underscored the seriousness of the 13 August clash by staging nationwide anti-Soviet demonstrations, as had been done in March. For their part, the Soviets concentrated largely on reportage from the scene of the incident while avoiding any effort to make use of the confrontation to develop a rationale for sterner measures or to condition the population for more serious developments. Moscow's campaign to rally nationalistic fervor, however, continued to have unsavory racist overtones, notably a *Pravda* military correspondent's dispatch depicting a Soviet soldier as having a golden tan, blue eyes, and golden locks while describing a Chinese prisoner as "a little man with brown eyes full of fear."[26]

If the summer of 1969 was dominated by border incidents that left the world bemused and apprehensive over the grave dangers latent in such a volatile situation, there were other signals of developments that would have transforming effects on the situation. One important source of these signals was the United States, which at this time began to propound a new policy. Most notably, President Nixon took the occasion of a press conference in Guam on 25 July en route to a tour of several Asian countries (plus Romania, a country with political lines of communication open to both Moscow and Peking) to enunciate a new doctrine for U.S. involvement in Asia. Known as the Nixon Doctrine, it called for Asian countries to assume a larger responsibility for their defense and regional security, with the United States providing a strategic shield and assistance in forms other than the direct commitment of American troops. Applied to the Vietnam War, this policy became known as "Vietnamization," a phased withdrawal of American troops and the concomitant assumption of the fighting burden by indigenous allied forces. It was also in mid-1969 that the first in a carefully graduated series of unilateral steps was taken by Washington to relax trade and travel restrictions governing mainland China, moves that were significant primarily in the signaling process.

Given the confrontationist atmosphere of the border crisis, it is not surprising that the Nixon Doctrine should have received a hostile notice by Peking. The Chinese gave their first public notice of the Nixon trip in a 27 July report on his visit to the Philippines.[27] Though not mentioning the Guam press conference, NCNA enunciated what became Peking's basic line on the Nixon Doctrine in its first two years by saying that the United States was trying to "resort more slyly to the counterrevolutionary tactics of using 'Asians to fight Asians'." Noting that the President said the United States would continue to fulfill its commitments in Asia (which to Peking meant, above all, U.S. commitments to the Nationalist Chinese regime on Taiwan), NCNA portrayed an attempt to "rig a new military alliance against China." Later, in a 5 August review of the Nixon tour, NCNA remarked that the "notorious god of war" had come to Asia under a banner of a new policy in an attempt to extricate the United States from its predicament by employing "new, counterrevolutionary double-dealing tactics." NCNA found "nothing new" and no change in the U.S. involvement in Asia apart from these more insidious tactics. What it did perceive was an analogy between Nixon's policy and Brezhnev's Asian collective security system.[28] Earlier, in Peking's first authoritative comment on Brezhnev's proposal, Chou En-lai on 13 July had denounced it as a new step by "social-imperialism" to create "a new anti-Chinese military alliance." Alluding to SEATO and CENTO, which had "fallen apart and exist only in name," Chou said the Soviets were "simply stepping into the shoes of U.S. imperialism."[29]

Chou's formulation was an interesting one, somewhat reminiscent of Chinese commentary several months earlier when Peking had made its overtures to the United States. This formulation also anticipated the differentiation between a receding U.S. involvement and an expanding Soviet one that would underlie and justify the basic shift in the Sino-American relationship of which Chou was the principal architect. Though, at this early stage, a complex signaling process had far to go, the United States was now beginning to provide the feedback that would nourish the Chouists as they proceeded to reconstitute the initiatives of change that had previously foundered.

Meanwhile, the view from Moscow was one of a hopeful prospect for improved Soviet-U.S. relations and of a bleak awareness of the depth of Chinese hostility. In a foreign policy statement to the Supreme Soviet on 10 July that included an extensive exposition of the Kremlin's China policy, Foreign Minister Gromyko[30] glumly observed that "even our most rabid enemies have never resorted to such unworthy methods and on such a scale" as had the Chinese in trying to discredit the Soviets. He anticipated what would become explicit Chinese policy in saying the Chinese perceived the Soviet Union as their main enemy. While firmly reasserting Moscow's position on the border question and warning that the border was inviolable across all its length, Gromyko renewed the proposal to resume the 1964 "consultations" and took favorable note of the fact that the joint commission was meeting on the "narrow" question of river navigation. He closed his discussion of China by saying Moscow had repeatedly stated its readiness to discuss economic, scientific, and cultural relations with China and to hold negotiations on "a large range of questions." Moscow was eager to defuse the border confrontation and to broaden its diplomatic options, a desire that would be nourished by suspicious signals between Washington and Peking auguring a change in the relationship between the Kremlin's two main adversaries.

Gromyko's policy statement sought to impart greater substance to Brezhnev's Asian collective security proposal by providing Moscow's first high-level elaboration of that nebulous idea. Above all, he sought to allay Asian suspicions, well founded, that this was a project aimed against China, dismissing as "absolutely groundless" the view that the proposal was directed "against some country or group of countries." Moscow clearly faced an uphill struggle, considering the growing aversion among Asians to becoming entangled in great-power polarizations, in its effort to convince the Asian countries that the Brezhnev proposal was not a Soviet version of John Foster Dulles's "pactomania" in the 1950s. Giving authoritative endorsement to diplomatic soundings then being taken in Asian capitals, Gromyko urged that consultations should be held among interested states at which the Soviet Union would offer "concrete considerations." It was undoubtedly to undercut Gromyko's overture that Chou three days later delivered his denunciation of the proposal as an attempt to create a new anti-China military alliance. The Chinese thus put the Asian countries on notice that the Soviet proposal meant further polarization, and this was something that was going out of fashion as a result of weariness from the Vietnam War.

Gromyko's statement contained signals to the Nixon Administration of a notably positive nature, which may have reflected a sense of confidence on Moscow's part regarding the state and prospects of East-West relations. In his conciliatory treatment of Soviet-U.S. relations, Gromyko raised the possibility of a summit meeting and of the commencement of strategic arms control talks, which had been twin casualties in the international arena of the invasion of Czechoslovakia. Moscow's confidence in mid-1969 must have been substantially bolstered by its success in June of holding the 75-party International Communist Conference, another casualty of Czechoslovakia that had quickly recovered. The very fact that the conference was finally held testified to the Kremlin's success in muting the controversy generated by the invasion, and this was particularly evidenced in the muffling of the Czechoslovak issue at the conference. In addition, though he was addressing what was billed as an anti-imperialist forum as leader of what was still widely acknowledged as the premier party in the Communist movement, Brezhnev's address to the conference was resonant with the language of East-West détente. This aspect of Brezhnev's message, as in the case of his proposal on Asian collective security, was amplified in Gromyko's policy statement in July. In addition to the indirect recognition that the damage to East-West relations from Czechoslovakia was now substantially repaired, Gromyko indicated that the Vietnam War was not a major obstacle to improving relations. In notable contrast to his policy statement to the Supreme Soviet in mid-1968, in which he said that relations with the United States were burdened by U.S. aggressiveness in Vietnam, Gromyko this time did not introduce the Vietnam question in the context of Soviet-U.S. relations. Rather, he chose to strike a positive note regarding the international implications of the Vietnam conflict, observing in another passage that an end to that war and a political settlement would "greatly contribute to the normalization of the international atmosphere as a whole." The United States having in 1968 put a cap on its rising involvement in Vietnam and also having opened negotiations on that question, and the Soviet Union in the meanwhile having put its own affairs (except for China) in order, the Kremlin was beginning to explore the options that a post-Vietnam settlement might offer.

The most significant signal contained in Gromyko's statement was his echo of what had been the keynote of the Nixon Administration's foreign policy, namely the President's call for an era of negotiation to replace one of confrontation. Gromyko's reference to that theme was the first by a Soviet leader, and it was important that this calculated feedback came in a major policy statement that was notable for seeking to open lines of communication and negotiation on the highest level and on the most basic of issues. Moscow was clearly intent on imparting new diplomatic momentum in a world scene that had been partially stalled by Czechoslovakia and then gravely complicated by the Sino-Soviet border crisis.

But, despite its forthcoming approach toward the West, Moscow's maneuverability on the world scene remained constricted by the confrontation with the Chinese. *A priori*, it might have seemed that both Moscow and Peking would have been stimulated by that confrontation to proceed further toward diplomatic activity with the United States—a prospect that had appeared hopeful in the period immediately before Czechoslovakia, in the case of Moscow, and immediately after the invasion, in the case of Peking. As it turned out, if paradoxically so, the opening of lines of communication to Washington from both Moscow and Peking was conditioned by a significant move toward defusing and controlling the Sino-Soviet border crisis. That crisis was symptomatic of underlying forces transforming the international environment, but the acutely enflamed symptoms needed to be brought under better control before new directions could be seriously explored.

Moscow's perception of the border confrontation as a political liability and a constraint on its diplomatic activity was given authoritative expression in an important article by Brezhnev assessing the results of the June international party conference. Published in the international movement's journal, *Problems of Peace and Socialism*, as well as in *Kommunist*, [31] the article drew satisfaction from the round of denunciations of Peking that took place at the conference, and it made the claim that the conference had "resolutely condemned" the ideology and policy of Maoism. But, while seeking to draw full ideological dividends from the conference investment, Brezhnev was notably restrained in dealing with the Sino-Soviet confrontation as a geopolitical fact of life. Especially significant was Brezhnev's characterization of the Soviet approach as one of "calm and restraint," and his assurance that Moscow would not be "provoked into thoughtless acts of any kind." There was no direct reference to the border question, though he did reaffirm Moscow's "firm and resolute" determination to defend the homeland. Brezhnev's message, which may have been intended for itchy-fingered leaders in the Soviet Union as much as for the Chinese or the international community, was that the Kremlin was determined to keep the border crisis in control and to avoid seizing on provocations or pretexts to take tougher action. In fact, to attempt to identify the primary audience for Brezhnev's message would be to distort this form of communication, for a signal of this kind acquires its full significance by virtue of the reinforcing effect of its being transmitted to several interacting audiences. For the chief of the CPSU to put his personal authority behind this message meant that more bloody-minded Soviet officials would have narrower scope for reacting militarily to the Chinese, and that the more flexible elements on the Chinese side would correspondingly have broader leverage for restraining the hard-liners in their midst. Gromyko's offer of a broad package of improved relations served a similar purpose, and indeed it was such a package that Kosygin would later seek to deliver directly to the Chinese in a surprise visit to Peking.

While holding out the prospect of opening lines of communication, Moscow at the same time took care to convey to the Chinese its determination to use force if necessary to cope with challenges along the border. Two anniversaries in August 1969 were pointedly relevant for conveying this warning. One was the 40th anniversary of fighting along the Sino-Soviet border in 1929, an occasion Moscow used to draw some ominous parallels to sober up the Chinese. Significantly, the Soviet military paper *Red Star* on 6 August carried an article by Col. Gen. V.P. Tolubko, formerly Deputy Commander of the Soviet Missile Forces and now Commander of the Far East Military District opposite China, marking the anniversary of the formation of a special Far East army in 1929 to deal with the Chinese along the border.[32] Tolubko said the Soviets had "tried to liquidate the conflict by peaceful means" and had attacked only after the Chinese invaded Soviet territory. He introduced an ominous note by recalling that Soviet units had remained on Manchurian territory "for some time," and he drew explicit parallels with the present situation in warning the "treacherous clique of Mao Tse-tung" that any attempt to "encroach on our motherland" would again be decisively rebuffed. A summary of the article broadcast to the Chinese highlighted the current situation and cited the "disastrous defeat" of the Chinese in 1929 to warn against any attempt to violate the Soviet border again.

Though Tolubko's current position and his background in missilery lent a certain calculated saber-rattling overtone to his minatory message, the thrust of that message was to warn the Chinese against aggressively challenging Soviet jurisdiction over contested border areas. In August 1967, when Marshal Yakubovskiy, speaking on the same occasion, announced to the troops of the Far East Military District that they were being awarded the Order of the Red Banner in succession to the earlier special army, the message was one of vigilance during a time of turmoil in China that affected Sino-Soviet relations generally, including border relations. Yakubovskiy had, however, focused on current dangers posed by U.S. actions in the Far East, in contrast to Tolubko's express concern to concentrate attention on the anniversary's implications for the current border dispute. The most significant contrast in this respect was that Yakubovskiy had mentioned only incidents in 1929 in which the Chinese had seized Soviet properties in Manchuria, a memory that had more relevance for 1967 (the year in which the Chinese besieged the Soviet embassy and detained a Soviet ship), whereas two years later Tolubko recalled that the Soviets attacked in 1929 after the Chinese had invaded Soviet territory in the Transbaykal and Primorye regions.

Still another anniversary in August, this one coming after the serious clash of 13 August in Central Asia, was used by Moscow to convey a stern warning to the Chinese. This was the 30th anniversary of the decisive battle of Khalkhin Gol in the 1938-1939 undeclared war between the Soviet Union and Japan in which Soviet and Mongolian troops delivered a telling blow to the Japanese. To mark the occasion, the Mongolian Defense Minister was in Moscow and a Soviet military delegation was in the Mongolian People's Republic to take part in commemorative ceremonies. A *Pravda* article on 19 August spelled out the current relevance of the anniversary by citing the Khalkhin Gol battle as "a decisive warning" to all who encroach on the MPR's independence, "including the Maoist adventurers," and quoting President Podgornyy as having said recently in Ulan Bator that no one must be in any doubt that any attempt to undermine the integrity of the Soviet or Mongolian borders would meet with a crushing rebuff.[33] The most authoritative message on the occasion was delivered, appropriately enough, by the military organ *Red Star*, which issued an editorial headed "The Borders of Our Motherland are Inviolable."[34] Drawing on nationalistic sentiment, the editorial began by invoking the names of Damanskiy (Chenpao) and Zhalanashkol (site of the 13 August clash) as having merged in Soviet consciousness as a symbol of the courage, gallantry, and heroism of "the sons of the socialist fatherland." Significantly, *Red Star* played on the historical overtones of the site of the 13 August clash, near the Dzungarian Gate, the traditional gateway in Central Asia between the Chinese and Russian empires whose centuries-old rivalry now seemed to have arrived at an explosive point. *Red Star* characterized this as an "ominous and symbolic" site evoking national memories of the hordes from the east that invaded Russia. But, while portraying a surge of anti-Soviet hysteria engulfing China in the wake of the 13 August incident, the editorial was also significant for its observation that there was "an enormous gap" separating China's "clamorous declarations" and its real potentialities as a threat to the Soviets. Thus, while sternly warning the Chinese against further enflaming an explosive situation, Moscow was avoiding any attempt to prepare the ground for justifying a preemptive attack against an adversary that might be portrayed as posing a serious threat to Soviet security. This form of negative signaling was as significant in the communications process as the strongly minatory messages also being conveyed. In certain contexts, the absence of certain signals can be as significant as the clamorous communications that naturally attract attention.

On the same day that the *Red Star* editorial appeared, the Chinese Foreign Ministry issued a note placing the 13 August clash against a background of continuous border incidents in June and July. The note was designed as a counterpoint to Moscow's professed desire for urgent measures to defuse the situation, and thus the note went to great lengths to offer a meticulous accounting of Soviet aggressiveness: in charging the Soviets with having repeatedly fired at Chinese islands and the Chinese bank of the border rivers, the note specified the figures of 1,116 bursts of fire and 943 single shots directed at Chenpao and the Chinese bank near the island. The note gave an overall count of 429 incidents in June and July in which the Soviets allegedly intruded into Chinese territory, fired on Chinese personnel and territory, obstructed Chinese border patrols, or interfered with Chinese vessels plying border rivers.[35]

The period after the 13 August clash saw an intensive Chinese polemical campaign against the "new tsars" in the Kremlin, a campaign fed by a torrent of material pegged to the anniversary of the invasion of Czechoslovakia as well as by the border dispute itself. This was particularly fitting, given the role played by Czechoslovakia in the genesis of the border confrontation in the past year. The border issue and Czechoslovakia were the subjects of documentary films released for showing on 20 August, obviously timed for maximum exploitation of the first anniversary of the invasion. Peking made a point of publicizing the showings of the films outside China in an effort to keep alive the memory of Czechoslovakia as part of the portrayal of Soviet social-imperialism—a portrayal which, as has been seen, was very much a part of the genesis of the border crisis.

It was in this setting—heightened border tensions after the serious 13 August clash, enflamed nationalistic feelings, and Peking's virulent polemical campaign—that Moscow made its most authoritative appeal to the international community for understanding of the perils of the crisis. Again the channel was a *Pravda* editorial article, on 28 August,[36] the latest in a series that included *inter alia* the one on 2 September 1964 marking an earlier serious phase of the border dispute and a landmark editorial article on 27 November 1966 opening Moscow's major offensive to isolate Mao "and his group" as illegitimate usurpers of a Communist regime. That the August 1969 article was a major communication was evidenced, not only by its format as an editorial article, but also by its being disseminated in full text by the TASS international service and being broadcast in full twice by Radio Moscow. In addition to being the most important Soviet pronouncement since the 13 August incident and being responsive to a massive polemical campaign, the editorial article was the first major Soviet policy statement on the border issue since the 13 June Government Statement proposing talks within two or three months. *Pravda* noted that the Chinese had not replied to the June proposals, and the fact that the three-month period set for the opening of talks was now about over was surely a factor in Moscow's decision to set forth the stakes involved in the crisis. Consistent with this timing factor, the editorial article was notable for its urgent warning against Peking's "dangerous, madly reckless position on matters of war and peace affecting the destinies of the peoples of the world." In developing this theme of the dangers to world peace posed by the Sino-Soviet confrontation, *Pravda* was amplifying and sharpening the line taken by Brezhnev in presenting Moscow's case at the June International Party Conference.

The *Pravda* editorial article was dictated as much by Moscow's awareness of the diplomatic constraints imposed by the border crisis as by a need to impress on the Chinese the dangers of their "adventurist" course. In its urgent appeal for international understanding, *Pravda* stressed that "war, should it break out, would be a terrible calamity for all mankind," and it called on all countries "to realize their responsibility to mankind" and seek ways to strengthen peace and security. This approach was consistent with Moscow's strategy in the Sino-Soviet rivalry to present the Soviet Union as a responsible member of the international community in contrast to a provocative and lawless China. One aspect of such an approach had been to seek to identify and encourage elements in the West responsive to an offer of détente, an approach that had borne fruit over the years in a series of East-West agreements violently opposed by the Chinese. Reflecting Moscow's concern over a tendency in the international community to take a neutral stance on the Sino-Soviet border dispute, *Pravda* took exception to what it chose to portray as Western press satisfaction over Sino-Soviet tensions. Significantly, however, it introduced a distinction between "the dark forces of imperialist reaction" and "sober-minded" representatives of capitalist leadership circles who expressed "grave concern over the threat to peace" posed by Peking's policies.[37]

As part of the effort to warn of the dangers of a major conflagration growing out of the border crisis, the *Pravda* article contained Moscow's most direct reference to Chinese nuclear capabilities in this context. *Pravda* pointedly observed that Chinese arsenals were "being filled with ever more new weapons" and that a war would involve "lethal weapons" and modern delivery systems that would "not leave a single continent unaffected." Though an appeal to world understanding and a portrayal of a looming Chinese nuclear threat would be expected to figure in a Soviet case for a pre-emptive or preventive strike, the thrust of *Pravda's* message was that a war would have disastrous consequences and that, with such a calamity in prospect, Moscow's diplomatic and political efforts deserved worldwide support. And, significantly, the editorial article echoed earlier Soviet comment in attributing Peking's anti-Soviet campaign to internal needs and in dismissing Chinese bellicosity as representing no real threat to Soviet security. *Pravda* noted, on the other hand, that Peking's recklessness and the atmosphere of war hysteria complicated "the entire international situation," causing alarm not only in bordering countries but in others too. *Pravda* renewed Moscow's offer to examine with the Chinese leadership the possibility of normalizing relations, and it concluded with a firm warning, addressed directly to Peking, that any attempt to speak to the Soviet Union with the language of weapons would be rebuffed. But, while warning against what *Pravda* termed "dangerous situations" being created on the border, Moscow did not use the editorial article as the kickoff for a campaign aimed at rallying support at home and abroad for stronger actions against the Chinese.

The same issue of *Pravda* recommended the anti-Chinese articles in the 27 August issue of *Literaturnaya Gazeta* as depicting the "present regime of terror and violence" in China. In addition to the torrent of polemics against China, *Literaturnaya Gazeta* also carried a rousing article by Ernst Henry on the anniversary of the outbreak of World War II in which that hyperbolist described "many indications" that a third world war could not be ruled out. Drawing on themes he developed in earlier articles in depicting the Peking leaders as dreaming of "a new pan-Asian Genghis Khan empire" and as preparing to join with the West in an encirclement of the Soviet Union, he warned that "a new and very dangerous, perhaps the most dangerous, hotbed of war today is arising" from this situation.

Thus, as the long hot summer was drawing to a close in a highly charged atmosphere, Moscow was intent on warning the world of the perils of a situation the Kremlin found burdensome and constraining. On 10 September, just short of the end of the three-month period in which Moscow's 13 June proposal had envisaged the opening of border talks, the Soviets released a collection of Chinese documents purportedly captured during border incidents and showing that the Chinese were "planning and staging the armed provocations." In what in effect was a rejoinder to the 19 August Chinese note meticulously detailing alleged Soviet violations of the border, Moscow explained in releasing the documents that Chinese violations occurred "almost every day," totaling 488 cases from June to mid-August. Recalling the 29 March and 13 June Soviet statements proposing resumption of "consultations" on the border question and urging the Chinese to take measures to avoid armed incidents, Moscow complained that the Chinese had delayed replying to the proposal on consultations and had failed to avoid incidents.[38] In a short time that situation was dramatically changed.

TEN

International Communist Politics

The border crisis was an inherently volatile and unstable condition; that is to say, it was symptomatic of the underlying changes taking place in the wake of Czechoslovakia, and either a new pattern of equilibrium evolved to resolve these acute symptoms or an uncontrollable disaster threatened to erupt. Indeed, that perilous prospect was the gravamen of Moscow's urgent appeal to the international community conveyed in the weighty *Pravda* editorial article of 28 August. *Pravda* had appealed to other countries "to realize their responsibility to mankind" in order to avoid the "terrible calamity" threatened by the border confrontation. As it happened, it was within the Communist movement itself that this responsibility was effectively taken up. And, ironically, it was Hanoi, whose reaction to Czechoslovakia had so provoked the Chinese, which provided the setting in which this sense of responsibility was communicated to the parties concerned. The occasion was the death of Ho Chi Minh, an event occurring exactly a year and a day after Chou En-lai had sternly (and forebodingly) lectured the North Vietnamese on the harsh imperatives of Peking's anti-Soviet animus in the wake of Czechoslovakia. Those imperatives were by no means to abate now; rather, they were to take a new direction which would lead to a realignment of global relationships that the North Vietnamese—and here another ironic twist—were to find bitterly distasteful. As a catalytic agent in international Communist politics, Hanoi found itself promoting trends having adverse effects on its own interests.

The death of Ho Chi Minh on 3 September 1969 provided the occasion for the North Vietnamese to draw on the immense prestige of their venerable leader to issue a solemn appeal for unity in the international Communist movement—an appeal, in effect, for a reduction of Sino-Soviet tensions at a time of acute crisis. It was also, in effect, a response to *Pravda's* 28 August appeal to a broader international jurisdiction for help in containing the border crisis. Ho and his lieutenants had been skillfully walking a tightrope of neutrality between Hanoi's two giant patrons ever since the direct U.S. involvement in the Vietnam combat in the middle of the decade had internationalized that war in a big-power context. Earlier, the Chinese challenge to Moscow's ideological and political leadership of the Communist movement had brought out the affinities among the East Asian Communist states, the North Vietnamese and North Koreans sharing with the Chinese a basic distrust toward Moscow's détente proclivities and a demand for more militant pressure against the U.S. presence in Asia. But the escalation of American combat operations in Vietnam during the second half of the decade necessitated a punctiliously correct neutrality on Hanoi's part. For one reason, the North Vietnamese now needed Soviet military assistance, particularly advanced weaponry to counteract the modern arms the United States was committing to the struggle. As it happened, the initiation of the American air campaign against North Vietnam coincided with the visit to Hanoi in February 1965 of Premier Kosygin, acting as an emissary of the new post-Khrushchev Kremlin leadership's line of "united action" in behalf of the Vietnamese comrades directly under fire from "U.S. imperialism." Whatever effect the Soviet overture to Hanoi (and to Pyongyang, which Kosygin also visited) might have had in other circumstances in seeking to reverse Khrushchev's loss of influence in East Asia, the Kremlin's offer of close relations was given real substance now in the form of arms aid, particularly air defense weaponry during a period in which U.S. air strikes were a central element in Washington's effort to thwart a Communist takeover in South Vietnam.

Another reason for Hanoi to adhere to a neutral position in the Sino-Soviet conflict proved to be more complex in practice. As the beleaguered "southeastern outpost of the socialist camp," North Vietnam was concerned to preserve the deterrent credibility of the Communist alliance system, and this called for at least containing and preferably ameliorating the bitter rivalry between Moscow and Peking that was calling into question the credibility of "socialist camp solidarity." Moscow skillfully played upon this demand of the times by combining its appeal for united action with a call for a polemical moratorium with the Chinese. Both of these complementary aspects of the post-Khrushchev Soviet line became issues in the Chinese strategic debate in 1965, with the result that the pressures of the Vietnam War became a divisive rather than a consolidating factor in the Sino-Soviet relationship when the prevailing Maoist wing of the Chinese leadership opted for an intransigent rejection of the appeal for Communist unity. Moreover, the Chinese stance not only complicated Hanoi's need for a show of socialist camp unity as a deterrent factor in American calculations, but Soviet material aid itself became an issue in the bitter rivalry between Peking and Moscow. The aid issue was a particularly sensitive one for Peking, which realized that the Soviets sought to use it as a wedge for reasserting their influence in East Asia. Peking was hardly eager to abet the Soviets in making themselves an indispensable patron of Hanoi. Moreover, by abetting Soviet aid transshipment via China, Peking would be helping Moscow to have its cake and eat it too: Moscow would be acting as Hanoi's aid patron while avoiding the threat of a confrontation with the United States arising from shipments by sea and a possible American blockade. This was the background against which Moscow bitterly accused the Chinese of obstructing Soviet aid deliveries, and Peking taunted the Soviets for not wishing to use the sea lanes for these deliveries.

Thus, Hanoi had had to exercise the greatest care not to antagonize one of its major patrons by leaning toward the other; and, in fact, the North Vietnamese had enjoyed considerable success in maintaining this balancing act and even in helping to control the Sino-Soviet conflict from spiraling completely out of hand. However, the border crisis of 1969 threatened just such a calamitous prospect, and this was the setting in which Hanoi sought to draw on Ho's legacy in order to play a part in crisis management. Shortly after his death, Hanoi declared a mourning period to run from 4 to 10 September, and on the 9th North Vietnamese party leader Le Duan publicly read Ho's last testament and delivered a eulogy.[1] In the testament Ho said:

> I am very proud to see the growth of the international communist and workers movement, but very grieved by the dissensions that are dividing the fraternal parties. I wish that our party will do its best to contribute effectively to the restoration of unity among the fraternal parties on the basis of Marxism-Leninism and proletarian internationalism, in a way consonant with the requirements of heart and reason.

Responding to this injunction, Le Duan concluded his eulogy by pledging to "constantly enhance Ho's pure internationalist sentiments" and to strive to contribute to restoring unity "in the socialist camp and among fraternal parties." Given the context of the Sino-Soviet border crisis, this was a message having very great current resonance.

Chinese behavior in the wake of Ho's death displayed some peculiar aspects, symptomatic of important developments behind the scenes. On the day after Ho's death, Chou En-lai hastened to Hanoi to offer condolences, beating his Soviet counterpart, Kosygin, by two days. However, Chou stayed in Hanoi only a matter of hours, and yet another delegation under Vice Premier Li Hsien-nien arrived in the DRV capital on 8 September to take part in the main memorial ceremony on the following day. A joint communiqué on the two Chinese delegations' visits, disseminated by the (North) Vietnam News Agency on the 10th, carefully obscured the fact of Chou's brief stay by lumping their activities together. Strangely, Peking failed to publish the communiqué. On the other hand, the Kosygin delegation, which departed from Moscow on the 5th, stayed in Hanoi from 6 to 10 September. A communiqué released by both sides quoted the Vietnamese leaders' pledge, in fulfillment of Ho's "behests and guided by reason and feeling," to seek to contribute effectively to "restoration of cohesion of the fraternal parties on the basis of Marxism-Leninism and proletarian internationalism." Seizing the opportunity, the North Vietnamese were actively lobbying for a reduction of Sino-Soviet tensions, and the Soviets were eager to provide positive feedback for Hanoi's signals.[2]

The pattern of events surrounding Ho's memorial services suggests that the Chinese were engaged in some significant decision-making. The dangers of the border crisis had been spelled out in chilling terms by *Pravda* on 28 August, and the impending expiration of the three months for the opening of border talks, as proposed by Moscow on 13 June, lent Hanoi's fervent appeal particular urgency. Interestingly, there was striking evidence even before Ho's death that the Chinese were already reaching a point of decision on how to deal with the crisis. An article in the ideological journal *Red Flag* for September, but already disseminated on NCNA's domestic service on 31 August,[3] raised issues that had figured in important Chinese leadership struggles in the past and had significant implications for the present situation. The article took the form (so very Chinese!) of a lengthy criticism of an opera, "The Beleaguered City," which had originally been produced in 1959 (the year of Defense Minister P'eng Te-huai's dismissal) and filmed in 1964, "when an extremely sharp class struggle was going on at home and abroad," as *Red Flag* pointed out. The opera deals with the battle of Changchun during the Chinese Civil War. According to the *Red Flag* article, the opera "negated revolutionary armed struggle" and exaggerated the "secondary contradictions" within the Kuomintang at the expense of the principal contradictions between the people and the enemy. While acknowledging that intelligence should be collected on conflicts and contradictions within enemy ranks, the article stressed that "what is most important is that we must base everything on our own forces and that *we must on no occasion be under any unrealistic illusion* about the contradictions within the enemy ranks" (emphasis added).

The sharply polemical cutting edge to this article signaled that it had important current implications; and indeed the article was not coy in identifying some of the vested interests involved. Thus, the article stressed that Changchun was liberated because the PLA, "directly commanded" by Lin Piao, had used armed struggle and had struggled against the "right-deviationist capitulationist line." The article also condemned the opera and other "poisonous weeds" of its time for having "advocated winning victory through negotiations" and attributing victory to the reasonableness of the enemy. At a "crucial" moment of class struggle, the article explained, Chiang Ch'ing (Mao's wife) had led a revolution in Peking opera to portray the merits of "the great people's army." Lin Piao as commander of the great people's army, Chiang Ch'ing as his ally in promoting the army's role: the *Red Flag* article had significantly identified two crucial figures in the Cultural Revolution, and it had raised issues that would necessarily figure in any departure by Peking from the dual confrontationist policy Lin had enunciated in his 1965 tract on people's war and which had dominated Peking's conduct of foreign policy during the Cultural Revolution.

The timing and polemical issues of the *Red Flag* article[4] provide a context for understanding Chou's hasty trip to Hanoi (a trip that consumed no more than the minimum time for paying last respects to a leader of Ho's stature): Chou's presence in Peking was required for the crucial decision-making then taking place. And what was soon to take place was a significant departure away from just that dual confrontationist policy (of Lin's) which the *Red Flag* article had polemically defended, and toward the negotiationist approach (of Chou's) which that article had decried. An early signal of this turning came when NCNA's international service on 10 September carried the text of Ho's testament, including the appeal for Communist unity. Ho might have been a sacred figure, but in recent years Peking had spared no one's sacrosanctness or sensitivity in its determination to censor any appeal for Chinese reconciliation with the Soviets. Now, however, Peking was signaling its (at least tentative) readiness to respond to Hanoi's mediatory efforts, not (as events would demonstrate) in order to return to Moscow's good graces but as a means of controlling the border crisis. As the *Red Flag* article angrily anticipated, the new Chinese approach would be based on "contradictions" within the enemy ranks to compensate with external leverage for what China lacked in relying on its own political and strategic resources. It was to position itself for this approach that Peking was now prepared to arrange with Moscow new rules of the game to control the border confrontation.

As usual, the Chinese decision-making process at this time had a wide context embracing internal relationships as well as more specifically foreign policy issues posed by the border crisis. Interestingly, there was a dearth during this period of new Mao "instructions," in contrast to the autumns of 1967 and 1968 which had been prime times for the issuance of new directives by Mao. Those earlier periods had been ones of consolidation after phases of upheaval. The absence now of Mao instructions may have reflected divided councils at a time when fundamental policy issues needed to be examined and possible new departures decided upon. As has been seen, the border crisis had served to immobilize Chinese policy on several fronts where new movement had been discernible in the wake of the invasion of Czechoslovakia. One area of relative immobility was that of party rebuilding, which had been proceeding painfully slowly since the 9th Party Congress in April had set the stage for reassertion of the party's institutional primacy after the ravages of the Cultural Revolution. Here was another area in which Lin Piao and his associates might well have perceived a threat to their interests from significant changes. Again the September *Red Flag* article can be seen as an important signal with its defensive insistence on the army's role and its recollection of Chiang Ch'ing's contribution to the army's "cultural revolution" in what proved to be a prelude to the full-scale Cultural Revolution of the late-1960s.[5]

All of this, then, formed the background for the surprising announcement on 11 September that Kosygin had met with Chou in Peking on his way back home from Ho's funeral. This meeting took place six days after Chou's hasty return to Peking from Hanoi to take part in the major deliberations that would issue in the invitation to Kosygin to make his first China visit since early 1965. Kosygin had, in fact, already reached Soviet Central Asia before doubling back to go to Peking.[6] That the Soviets would have their Premier go to such lengths, reminiscent of the traditional pattern of foreign emissaries having to go to Peking to meet with leaders of the Middle Kingdom, demonstrated their desire to bring the border crisis under control. This desire was also reflected in the positive cast lent to Moscow's announcement of the Peking meeting. According to Moscow, a meeting was held "by mutual agreement" at which the two sides "candidly explained their views, holding a conversation useful to both sides."[7] By normal standards of relations between Communist states, that characterization was quite frigid, but by comparison with Peking's treatment of the meeting the Soviet version was almost effusive. Peking's report, coming over five hours after Moscow's, noted that the meeting was held at the airport (a slight to Kosygin not mentioned by Moscow) and tersely recorded that the two Premiers had "a frank conversation," with no indication that the meeting was useful or that the Chinese were doing anything more than tolerating a brief stopover.[8]

Kosygin's journey to Peking having dramatically demonstrated Moscow's desire to defuse the border crisis, the Soviets proceeded to reinforce that signal and to minimize distortions from conflicting signals. In particular, the Soviets began a polemical standdown, a move designed to demonstrate Moscow's *bona fides* in seeking to move the border issue onto the negotiating table.[9] This move to clear the communications channels was undoubtedly meant to reinforce those elements in the Chinese leadership counseling a containment of the border crisis; conflicting signals from Moscow would be seized upon by hard-line elements eager to exploit ambiguities in Moscow's stance. Moscow's awareness of this aspect of the communication process was especially reflected in a statement by Soviet Foreign Ministry spokesman Zamyatin, as quoted by TASS on 23 September,[10] denying that

Moscow had information on Mao's rumored illness and charging that Western press reports citing Soviet sources were "provocative rumors." Moscow was thus seeking to show that its hands were clean of any speculation on such sensitive matters as Mao's health and its implications for Chinese decision-making. Beginning on 19 September, a day after the Chinese sent a secret message to Moscow on opening border negotiations, Soviet broadcasts to China began to make references to the propaganda stand-down, saying that world public opinion regarded the cessation of polemics as a new initiative by Moscow aimed at normalizing Sino-Soviet relations. Also on the 19th, Moscow had recourse to the mode of indirect communication through recalling past troubles on the border. An article in *Sovetskaya Rossiya*, reporting on an ongoing publication of Soviet foreign policy documents, discussed Chiang Kai-shek's seizure in 1929 of the Chinese Eastern Railway and related border incidents.[11] The article noted that Moscow had "repeatedly" proposed a negotiated settlement but that "the Soviet command had to take the necessary measures to defend our borders" after the proposals had been rejected. "Ultimately," the article added, Chiang agreed to create a joint commission to settle the dispute peacefully. The article thus implied a limit to Moscow's forbearance, but the reference to a negotiated settlement introduced a new element to the historical analogies Moscow had been drawing that summer.

During this crucial period, Moscow was also seeking to reinforce its signals to Peking by drawing on the sentiments of other elements of the Communist movement. Hanoi was being particularly active in the wake of Ho's death; an editorial in the party daily *Nhan Dan* on 15 September, for example, highlighted the lament in Ho's last testament over discord in the Communist movement and pursued its message: "Whether in Vietnam or anywhere else," the editorial asked plaintively, "is there any communist, any revolutionary fighting under the banner of Marxism-Leninism, who is not impressed by these heartfelt words" of Ho's? Two days later, on the 17th, another *Nhan Dan* editorial devoted to the state of the international Communist movement delivered a message Moscow could tailor to its own signaling. In a front-page summary of the editorial, *Pravda* quoted *Nhan Dan* as warning that the imperialist enemies were undertaking "subversive splitting activities," and emphasizing that the entire socialist camp had "a common, very dangerous enemy in imperialism, headed by American imperialism." *Pravda* quoted the editorial as appealing against this background for the "restoration, consolidation and development of the solidarity and unity of the revolutionary forces throughout the world in accordance with the principles of proletarian internationalism."[12]

That it was Moscow that was taking the lead in trying to defuse the crisis was also indicated by Peking's failure for some time to reciprocate the Soviet signals following the Kosygin-Chou meeting. Chinese polemics against the Soviets at first continued unabated. The border crisis was very much in evidence in the set of slogans issued by Peking on 16 September for the 20th anniversary of the PRC to be marked a fortnight later. One slogan, later to be identified as a Mao directive,[13] called attention to the possibility of a Soviet nuclear attack. The slogan called on the "people of all countries" to unite and oppose "any war of aggression launched by imperialism or social-imperialism, especially a war of aggression in which atom bombs are used as weapons. The people of the world should use revolutionary war to eliminate the war of aggression, should such a war break out, and preparations must be made against it right now!" This was a sharpened restatement of Lin Piao's formulation in his 9th Congress report concerning the possibility of a war being launched by the United States "and" the Soviet Union. The slogan's appeal to the people of the world to wage "revolutionary war" to counter Soviet aggression was a Maoist retort to the 28 August *Pravda* editorial article's appeal to the international community for support against a bellicose and irresponsible China. But more important contextual factors were the fact of the Kosygin-Chou meeting in the meantime and the Chinese message two days later—surely already in the making—probing the possibility of border talks.

Events surrounding the PRC's 20th anniversary, on 1 October, signaled the changes in Chinese policy now taking place. In keeping with the recent pattern, the anniversary ceremonies consisted of a reception hosted by Chou, a Peking rally attended by Mao and addressed by Lin Piao, and a joint editorial marking the event. Foreign Communist representation testified to Chinese readiness to make some accommodation in the interest of easing the border confrontation. A high-level North Vietnamese delegation was headed by Premier Pham Van Dong; in 1968 there had been no official DRV delegation, and Vietnamese representatives were of such a low level that they were listed below a New Zealand delegation. Even more revealing was the arrival of a North Korean delegation led by the DPRK head of state, who did not manage to appear in time for Chou's anniversary-eve reception and who was accorded a welcome interestingly different from that for the other delegations. Where the other delegations, including the Vietnamese, were greeted with slogans that included one urging "Down with social-imperialism," the welcoming chorus for the North Koreans changed that slogan to "Down with modern revisionism." This change was undoubtedly a Chinese concession to the North Koreans, who had long shared Peking's revulsion against Moscow's revisionist proclivities (and were not at all shy about saying so) but who objected to the conversion of the ideological rivalry into a military confrontation, as Peking's use of the term social-imperialism implied.[14] The variant slogan and the DPRK President's late arrival suggested a last-minute Chinese concession, one the North Koreans were shortly to reciprocate with positive feedback. On 8 October, after the Chinese belatedly announced two nuclear tests (see below), Pyongyang sent a congratulatory message signed by Kim Il-song in his party capacity and addressed to Mao as well as to Lin and Chou; only a few days earlier, the North Korean message on the PRC's National Day had been signed by Kim in his Prime Minister's role only and addressed only to Chou. Pyongyang thus signaled that it was renewing normal party relations with Peking, ending a period of estrangement during which the North Koreans voiced strong political and ideological objections to Chinese policy in the international Communist movement. The revival of the pre-Cultural Revolution affinities between the two neighboring Communist states would receive dramatic endorsement the following April when Chou led a delegation to North Korea to celebrate their common interests, a visit that underscored Chou's success in exercising the new flexibility in Chinese policy.

A more important signal of Chinese flexibility involved one of the National Day slogans issued on 16 September but revised by the time of the anniversary itself. In its original form, the slogan urged "Down with Soviet revisionist social-imperialism," but, in the anniversary speeches and joint editorial, the direct reference to "Soviet" was diplomatically omitted. Against the background of the Kosygin-Chou meeting and Moscow's polemical stand-down, this signal must have been heartening to the Soviets. Moreover, the joint editorial declared that Peking stood for the settlement of border conflicts through negotiations, the first Chinese reference to negotiations since the Kosygin-Chou talk.[15] To be sure, the Chinese also used the occasion to react to Soviet threats and pressures, the National Day editorial dismissing as "simply day-dreaming and madness" any attempt to instigate "remnant counterrevolutionary forces," and taking note of "futile attempts to organize rebellions" among minority nationalities along the border. The editorial also warned that any attack on China would plunge the aggressors into "the escape-proof net of a great, just people's war," a reminder of the bottomless pit to which Soviet military moves into China could lead. But it was left mainly to Chou, the principal Chinese negotiator, to deliver the diplomatic message. In his 30 September speech, Chou projected a relatively conciliatory approach, reproaching the Soviet Union (anonymously termed "social-imperialism") for its "war threats" but taking pains to depict Peking's policies as consistently peaceful. He followed a defense of China's nuclear weapons program as being solely for defense with a reaffirmation of Peking's adherence to peaceful coexistence. He added, in an allusion to the still secret correspondence on opening border talks, that the Chinese would "never barter away principles." He also alluded to the 28 August *Pravda* article in complaining that the Chinese had been accused of having expansionist ambitions and of intending to launch a nuclear war. As part of his effort to revive his old diplomatic image, Chou personalized a reference to peaceful coexistence when (speaking at a dinner for a Cambodian delegation in Peking for the anniversary) he harked back to his acquaintance with Prince Sihanouk at Bandung, the very symbol of a flexible and reasonable phase of Chinese diplomacy.[16]

In another indication of this new flexibility, Peking departed from its past practice of promptly announcing nuclear tests when it belatedly reported on 4 October that China had conducted an underground test on 23 September and a hydrogen bomb explosion on the 29th.[17] The announcement did follow past practice, however, in emphasizing the defensive nature of Chinese nuclear weaponry and repeating the pledge that China would not be the first to use these weapons. On the same day as this announcement, NCNA released an article in the October issue of *Red Flag* that took proud note of the rapid advances made by the Chinese in developing nuclear weaponry (in the process recalling Soviet efforts to inhibit the Chinese program) and predicted that the Chinese would scale the heights of world science and technology.[18] Most notably, the article cited Mao as having given importance to the role of intellectuals—one of the main targets of the Cultural Revolution—and as being greatly concerned over the healthy growth of intellectuals. The article signaled a move toward expertness at the expense of the ideological "redness" in fashion during the Cultural Revolution. Interestingly, in this connection, Lin Piao in his National Day rally speech said he was speaking in behalf of Mao, the party's Central Committee, and the government.[19] On the two previous National Day anniversaries, he had additionally cited the Cultural Revolution Group and the party's Military Affairs Committee, two organs that had played central roles in conducting the Cultural Revolution. The expert hand of Chou En-lai the diplomat was being visibly strengthened at the expense of the red baton with which Lin Piao had conducted the dithyrambic Maoist litanies of the Cultural Revolution.

Meanwhile, Moscow by its own signals was keeping the diplomatic option open for the Chinese, as reflected in particular by the Soviet message on PRC National Day.[20] For one thing, the message was relatively lengthy in comparison with the frosty brevity of the last few years. Significantly, the message called for "normalization of relations between our two *states*" and the settlement of conflicts "by peaceful means, through *negotiations and consultations*" (emphasis added). The reference to negotiations as well as consultations reflected an important concession by Moscow, which theretofore had adamantly insisted on limiting border discussions to "consultations" on specific boundary rectifications and thus avoiding the impression of opening the postwar Soviet borders to renegotiation. This was surely a necessary concession in order to bring the Chinese to the bargaining table, and it was further evidence of Moscow's strong interest in defusing the border crisis. The call for talks was reinforced by the proposition conveyed in the Soviet message that the fundamental interests of the Soviet and Chinese peoples coincide and that peaceful settlements of outstanding problems between the two states would contribute to the struggle against imperialism and to peace and security in the Far East and worldwide. The messages on this occasion in the three previous years had limited their treatment of Sino-Soviet relations to avowals that the Soviets would pursue a policy of "strengthening friendship with the Chinese people," a hope so vague and vacuous as to be a frank acknowledgment that relations between the two countries were almost totally immobilized on the diplomatic front. Now that immobility had been broken by movement toward the negotiating table and the attendant defusing of the border crisis.

ELEVEN

Onto the Negotiating Track

The climate of Sino-Soviet relations having become more propitious by the beginning of October 1969, a Chinese Government Statement on 7 October announced that agreement had been reached to open negotiations on the border conflict, a move that would serve to defuse the crisis without, however, marking a breakthrough in resolving the border dispute as such.[1] Conveying this purpose, the Chinese statement placed primary emphasis on measures to contain border tensions, urging military disengagement from disputed areas and, in effect, putting aside the more basic issues for which major concessions would be required for a settlement. Making a conciliatory gesture reciprocating Moscow's positive signals, the statement stressed that the Chinese were not demanding the return of the territory annexed by means of the "unequal treaties" and that there was "no reason whatsoever" for China and the Soviet Union to fight a war over the border question. The statement did cite the call in the PRC's 24 May statement for an "overall" settlement and for "all-round negotiations" on the border issue, but it failed to renew the demand that Moscow acknowledge the unequal nature of the nineteenth-century treaties, a precondition that Moscow would regard as intolerable, given its sensitivity to anything calling into question the validity of existing borders. Thus, in this delicate signaling process, Moscow had acceded to the opening of "negotiations" as well as "consultations," while Peking in turn had scaled down its demands.

In Peking's first reference to the Kosygin-Chou meeting since the original terse announcement, the Chinese statement revealed that the two premiers had had "an exchange of views" not only on the border question but also on trade "and other questions." The statement depicted Peking as having taken the initiative in proposing measures to control border troubles by saying the Chinese had proposed that the troops of the two sides disengage by "withdrawing from, or refraining from entering, all the disputed areas," as indicated by the maps exchanged during the 1964 discussions. According to the statement, the Chinese proposed that the two sides "first of all reach an agreement on the provisional measures" for disengagement and maintenance of the status quo. The 24 May statement had called for maintenance of the status quo on Chinese terms, in calling for each side to remain behind the line of actual control but qualifying this with the stipulation that troops should not cross the central line of the main channel of border rivers, or the thalweg. The 7 October statement, however, made no mention of the thalweg, a central issue in the conflict that had erupted in March; instead the statement referred to the two sides' 1964 claims as defining the disputed areas and put the focus on disengagement. The statement disclosed that the Chinese proposals had been forwarded to Moscow in secret messages on 18 September and 6 October, a time frame accounting for Moscow's feedback signaling in its message on PRC National Day.

While the 24 May statement and other Chinese pronouncements had referred to Soviet "nuclear blackmail" in the context of the border dispute, the 7 October statement was exceptionally explicit in taking note of speculation regarding a Soviet preventive strike on Chinese nuclear facilities. After reiterating the standard position that the Chinese would under no circumstances be the first to use nuclear weapons, the statement said China would never be "intimidated by war threats, including nuclear war threats," and added: "Should a handful of war maniacs dare to raid China's strategic sites in defiance of world condemnation, that will be war, that will be aggression, and the 700 million Chinese people will rise up in resistance and use revolutionary war to eliminate the war of aggression." This striking warning was designed to embarrass and isolate the intransigent hard-liners in the Soviet leadership who may have been contemplating such extreme measures; it also reminded the Soviets of the bottomless pit to which even a "surgical" strike could lead. Moreover, the Chinese having just recently conducted two nuclear tests including their first underground explosion, the stark reference to a Soviet strike put the spotlight on Moscow as the party that might be inclined to use military force rather than diplomatic means to resolve the border crisis.

In addition to agreement to hold border "negotiations," the Chinese statement disclosed a related concession in which Moscow pulled back from its previous terms. It was now agreed that the talks would be held at the level of deputy foreign ministers, whereas Moscow had previously proposed having its delegation to "consultations" be headed by the border guards commander. The implication of this pair of changes was that Moscow was acknowledging that there was a political issue requiring negotiation, not simply a technical question of clarifying the boundary line along some pieces of the border. Nonetheless, the Chinese emphasis was on provisional measures to contain border tensions and thus to defuse the crisis situation, and to serve that purpose the statement made a distinction between normalization of state relations and fundamental political and ideological differences between Moscow and Peking. Regarding these "irreconcilable differences of principle," the Chinese pledged a struggle persisting for "a long period of time." But this "struggle of principle," the statement explained, "should not prevent China and the Soviet Union from maintaining normal state relations on the basis of the five principles of peaceful coexistence." This was an ingenious formulation, for it served to justify the current moves to defuse the crisis while at the same time providing the ideological basis for conducting the long-term struggle that the Chinese challenges on the border issue had been designed to promote. By placing Sino-Soviet relations on the basis of peaceful coexistence between states, Peking was carefully separating the ideological and state levels of policy, thus facilitating a more flexible activity on the diplomatic front and freeing Chouist diplomacy from the encumbrances of ideological constraints. Moreover, this formulation implicitly rejected Moscow's orthodox doctrine that peaceful coexistence governs only relations between Communist and non-Communist countries ("proletarian internationalism" governing relations among Communist states and parties), a doctrine serving Moscow's interest in a bipolar international system in which the Kremlin attends to the interests of the socialist camp in its relations with the other camp. As has been seen, it was to challenge just such a bipolar system that Peking chose to pose a border challenge in the wake of Czechoslovakia.

Peking's new strategy was further delineated on 8 October with the release by the Foreign Ministry of a document offering a detailed rebuttal of the 13 June Soviet Government Statement proposing border talks.[2] That it was the previous day's statement that contained the operative message was indicated by its having been a Chinese Government Statement, the same level as the Soviet document of 13 June. Its message was that border talks could now open in the interests of provisionally regulating tensions, producing a new set of border practices while the stage was being set for a broader diplomatic and political contest over the long term. The 8 October document, on the other hand, was intended mainly for the record, setting forth again the detailed elements of Peking's position on the border dispute per se, and on this matter it was made clear that the Chinese position remained firm and unyielding. Apart from still another tendentious examination of the historical background, the new document took issue with the 13 June Soviet Statement for having expressed "in an equivocal way" Moscow's readiness to take the inherited treaties as the basis for a settlement. The document cited two examples to question Moscow's sincerity on this fundamental issue. First was the matter of the disputed 20,000 square kilometers in the Pamir Mountains, the Chinese arguing that they had made explicit reservations on the legal status of this area at the time of agreeing in 1894 to maintaining the line of actual control after the Russians had seized it in violation of the 1884 protocol on the boundary in the Kashgar region. Second, regarding the boundary along the Amur and Ussuri rivers, the Chinese cited an article in regulations on safeguarding the Soviet frontier ratified by the Supreme Soviet in 1960 stipulating that the boundary on border rivers runs along the thalweg. While acknowledging, in reference to Moscow's citation of the 1858 treaty between Costa Rica and Nicaragua defining their boundary as running along one river bank, that there are exceptions to established principles of international law such as the thalweg principle, the Chinese document argued that explicit stipulations must be made in treaties for any exceptional case, as in the 1858 Central American instance, but that no such stipulation could be found in the 1860 Treaty of Peking.

As these two points illustrated, Peking's position on "an overall settlement" hardly augured a breakthrough in the impending negotiations. As set forth in the 8 October document, the Chinese demands were that: (1) the unequal nature of the existing treaties must be acknowledged; (2) these treaties should serve as the basis for an overall settlement, in which China would not require the return of territory annexed by means of these treaties; (3) any side occupying the territory of the other side in violation of these treaties "must, in principle, return it unconditionally," though necessary adjustments could be made by mutual accommodation; (4) "a new equal Sino-Soviet treaty" should be concluded to replace "the old unequal Sino-Russian treaties"; and (5) pending an overall settlement, the status quo should be maintained and military disengagement effected by withdrawing troops from disputed areas as indicated by the differing delineations on the maps exchanged during the 1964 talks. It was the disengagement proposal that was at the focus of the Chinese statement of the previous day, and it would soon become clear that this issue alone would prove virtually as intractable as the entire package of issues constituting the border dispute. This was not surprising, in view of the severity of the clashes over control of otherwise insignificant Chenpao. Actual control being an important factor in territorial bargaining, one side, in this case the Soviet, was not likely to surrender something in the guise of provisional measures that would prejudice its claims in the bargaining process before it was really underway. It should be noted that the Chinese complained that, in previous talks, the Soviets had claimed over 600 of 700 islands on the Chinese side of the thalweg, islands also claimed by Peking. Using the criterion of disputed areas as defining the extent of disengagement, this would mean an extensive Soviet withdrawal even before any bargaining was underway. Not surprisingly, Moscow regarded these disengagement demands as totally unacceptable preconditions to negotiation.

Peking saw the border talks as a means of establishing a new equilibrium after the perilously destabilizing disturbances resulting from the Chinese challenges to the old rules of the game in the past year or so. For its part, Moscow, having chafed under the diplomatic and political constraints posed by the border confrontation, welcomed the border talks as an impetus to what it hoped would be an expanding normalization process. Differing perceptions and expectations were reflected in the two sides' treatment of the opening of the talks, which began on 20 October. Peking announced that the Soviet delegation, headed by Deputy Foreign Minister V.V. Kuznetsov, arrived on the 19th for "negotiations on the Sino-Soviet boundary question," thus indicating the narrow scope conferred by Peking on the talks.[3] Moscow's report on Kuznetsov's arrival in the Chinese capital said he came to conduct "negotiations on questions in which both sides are interested"—a formulation reflecting both Moscow's wariness about acknowledging a border question open to renegotiation and its interest in a broad agenda in order to engage the Chinese in a wide-ranging normalization process.[4] Moscow sought to project a propitious atmosphere by quoting Kuznetsov as expressing hope that the talks would be fruitful and as thanking his hosts for a cordial welcome; the Chinese chose not to offer feedback in kind. In another gesture that coincided with Kuznetsov's arrival, the set of slogans issued to mark the upcoming Bolshevik Revolution anniversary omitted a salute to the "valiant border guards" defending the homeland's "sacred borders" that had been introduced in the comparable May Day slogans issued in the wake of the Chenpao clashes.

Also during this time, with the world directing its attention to the opening of Sino-Soviet negotiations after a summer of acute tension, the North Vietnamese pursued their mediatory role, encouraging the two sides to defuse the confrontation. This was especially evident in remarks by DRV Premier Pham Van Dong in Peking on 23 October, three days after the opening of the border talks, while he was on the return leg of a visit to China, winding up a tour that included the Soviet Union and East Germany. The Vietnamese expressed "deep hope and wishes" for favorable results from the Sino-Soviet negotiations, which he termed of "important significance." Peking carried the text of his speech,[5] and Moscow even more pointedly provided feedback in a TASS dispatch the next day singling out the passage on the opening of the talks.[6] In a minor signal illustrating Peking's return to a more diplomatic approach after Cultural Revolution zealotry, NCNA's report on Pham Van Dong's return to Peking on 21 October noted that he was arriving from the Soviet Union and East Germany, a departure from Peking's practice in recent years of omitting to mention that an arriving delegation had also been consorting with the revisionist heretics. The message in all of this was that now Peking was not permitting ideological purity to stifle diplomatic flexibility.

For its part, Moscow was notably forthcoming in signaling its good will and its hopes invested in the negotiations. In the first authoritative Soviet comment on the opening of the talks, Brezhnev on 27 October[7] extended a conciliatory appeal for "a positive, realistic approach" at the negotiations, calling on Peking to match Soviet good will in order to settle their differences. He put his personal authority behind Moscow's readiness to negotiate, and again associated himself with counsels of moderation on the border confrontation in prefacing his comments on the talks with a reference to "our presence of mind and restraint" during a period of the most enflamed tensions. Brezhnev also signaled Moscow's interest in a broad agenda, saying the talks should cover "border and other questions." But the most notable signal came when, in recalling the Kosygin-Chou meeting in September, Brezhnev graced his reference to the Chinese Premier by calling him "Comrade Chou En-lai," a striking fraternal gesture that belied the absence of party ties and revived a practice of Communist protocol long dormant in Sino-Soviet relations.[8] Whatever the Soviets may or may not have known about the state of play of Chinese factional politics at this time, Brezhnev was authoritatively signaling an interest in initiating a serious negotiating process, and Chou seemed to represent those interests served by a negotiationist approach rather than the politics of confrontation. What Moscow could not foresee was that Chou was playing for higher stakes in a more complex game than an amelioration of Sino-Soviet relations could serve.

Though the new Chinese approach would, in the near future, take a turn that would bedevil the Soviets in complex new ways, and would infuriate the North Vietnamese who had applauded its origins, the immediate effect—and a very significant one—of the agreement to move to the negotiating table was to introduce a new stability in what had become a dangerously volatile situation. The crisis was contained, with the concomitant removal of the uncertainties and foreshortened perspectives that had paralyzed movement on various fronts. A year earlier, another crisis, Czechoslovakia, had been surmounted, and in his 27 October speech Brezhnev also assessed the meaning of that experience. The occasion was the presence of Czechoslovakia's Husak, who was in Moscow to receive Brezhnev's benediction in the wake of a Czechoslovak Party Plenum the previous month that purged many figures remaining from the Dubcek leadership and officially repudiated the party's resolution condemning the August 1968 invasion. Brezhnev found it timely to elaborate on the global significance of the Czechoslovak crisis, "now that the complex network of causes and consequences" of the crisis had "to a large extent been unraveled," and more and more people understood that "what happened in Czechoslovakia was one of the tensest postwar class skirmishes between the world of socialism on the one hand and international reaction and its agents on the other." In what Peking would have perceived as the systemic mirror image of the official U.S. interpretation of the Vietnam involvement, Brezhnev declared that many things were put to the test in the Czechoslovak crisis: the stability of socialism in Czechoslovakia and of the socialist system in Central Europe in general, and the impact of socialist internationalism on the solidarity of the fraternal countries. In systemic terms, he was saying that Soviet intervention in Czechoslovakia had helped preserve the bipolar character of the international system by redeeming Moscow's commitment to the system's stability. Reinforcing the warning signals from the Czechoslovak experience, Brezhnev cited both Hungary in 1956 and Czechoslovakia in 1968 in stressing the importance of implacable vigilance against "the revisionist and right-opportunist danger."

There were still other new elements in the international environment that Moscow perceived as conducive to new initiatives and longer perspectives than a crisis atmosphere would permit. In the same 27 October speech, Brezhnev scarcely concealed Moscow's satisfaction over the outcome of the recent West German elections, in which the Christian Democrats ("the party of big monopoly capital") lost power after 20 years. The signaling process on the German question was quickly initiated, as in Brezhnev's cautious but positive observation that the victorious coalition headed by Social Democrat Willy Brandt wished to take "a more realistic" approach in international affairs. Brezhnev seized the occasion to invite the new Bonn government to move to a policy line "corresponding to the real state of affairs on the European continent," a move that was promised "understanding and support" from Moscow. With what proved to be justified hopefulness, the Kremlin was looking toward a prospect in which the postwar division of Europe would be legitimized and its boundaries consolidated. Military intervention in Czechoslovakia having been one means to stabilize the postwar settlement, Moscow now felt itself in position to try the diplomatic instrument to further the same cause.

The negotiating front found yet another important opening in this post-crisis situation, or, more precisely, a reopening of what had been closed by the invasion of Czechoslovakia. On 25 October Moscow and Washington announced that preliminary discussions on strategic arms limitation would open the next month in Helsinki, the first step toward a series of arms control agreements. Interestingly, the agreement to open talks was arranged by Soviet Ambassador Dobrynin during a visit to President Nixon on 20 October, the very day the Sino-Soviet border talks opened. Washington had withdrawn from a previous agreement to open SALT a year earlier, in protest against the invasion of Czechoslovakia. The border crisis had subsequently immobilized Soviet diplomacy, but now, with that crisis defused, Moscow was prepared to accept an American proposal to initiate talks in parallel with the Peking talks.

The new post-crisis atmosphere was evident in Moscow's annual celebration of the Bolshevik Revolution anniversary in November. Thus, for the first time in four acrimonious years, the keynote speaker refrained from attacking the Chinese, choosing rather to use the occasion to express hope for fruitful results from the Peking talks. In noting that relations with the PRC had "sharply deteriorated" in recent years, the keynoter, President Podgornyy, conspicuously skirted the question of culpability by observing blandly that this had happened "for reasons you all know." The occasion was also used to signal the Kremlin's interest in constructive negotiations with the Brandt Government and with the United States on SALT. In another context, but dealing with a matter that would figure centrally in Moscow's détente policies, Podgornyy offered a notably candid acknowledgment of the regime's disappointment over an inadequate rise in the country's standard of living. He expressed particular concern that labor productivity had increased more sluggishly than expected, and here he dwelt on the importance of technological modernization as the focus of competition between the two rival systems. Access to Western technology was becoming a key element in the politics and practice of détente.[9]

If Moscow's foreign policy had become energized by the release from crisis tensions, the changes that were to ensue for the Chinese were even more fundamental and far-reaching in their implications. For Moscow, both the Czechoslovak and border crises were challenges to the international system, and the Soviet responses were efforts to restabilize the system. As the German question illustrated, the regained stability afforded Moscow scope to pursue a line of evolution looking toward further confirmation of the postwar international system. For Peking, as has been seen, Czechoslovakia and the border issue were related in a more complex way, the second having been a central instrument in the Chinese challenge to the existing system in the wake of the Soviet intervention in Czechoslovakia to preserve that system. It will be recalled that, after Czechoslovakia, the Chinese had pressed the border issue as part of a series of initiatives having important implications across the full breadth of foreign and domestic policy. The enflamed border issue had been symptomatic of the challenge being posed to the international system, but the new equilibrium on this issue achieved in the autumn of 1969 meant not an end to that challenge but a renewal of those trends in Chinese policy that could be discerned a year earlier. The border issue itself was now to be kept under control, its function having been served (perhaps overly well), and now the other trends found scope for their own development.

Two events occurring shortly after the opening of border talks signaled the directions now being taken by Peking. Chinese treatment of the Bolshevik Revolution anniversary in early November reflected the interest in defusing tensions and promoting "peaceful coexistence" between the two states. Having snubbed Moscow the previous year by failing to send a message on the occasion, Peking this time extended "warm greetings to the fraternal Soviet people" and held a reception hosted by the Chinese-Soviet Friendship Association. The Soviet embassy's traditional reception on the anniversary, 7 November, was attended by Deputy Foreign Minister Ch'iao Kuan-hua, head of the Chinese delegation at the border talks. It should be noted that all of these were symbols of more normal *state* relations, with no hint of restored party ties, and that, at the same time, the counterpoint to peaceful coexistence—the struggle over "irreconcilable differences of principle" promised by Peking—was given demonstrative expression on the very same day when the Chinese Politburo gave a reception for the ambassador of Albania, Peking's ideological ally, attended by a powerful turnout of leaders headed by Chou En-lai. This dual-track approach was to become characteristic of Chinese policy in the next decade.

The other indicative event at this time was the announcement of an agreement to restore diplomatic relations between China and Yugoslavia after an interval of 11 years. This represented a fruition of the new East Europe policy initiated by Peking in the wake of Czechoslovakia, and its appearance now dramatized the new flexibility and range of Chinese policy in the post-crisis atmosphere. Eleven years earlier, Sino-Yugoslav relations had become envenomed as an acute symptom of Peking's militant insistence at that time on a sharply bipolar international system while Belgrade sought to circumvent bloc politics. Fittingly, Sino-Yugoslav relations were now being repaired as an outcome of Peking's determined effort to challenge the bipolar system.

Apart from an evolving East Europe policy, another subject on which Peking had been seeking new directions in the latter part of 1968, the Vietnam question, was also now being given a fresh look by the Chinese. Though, at this time, it was at the stage of early warning signals, the Vietnam issue would provide a concentrated focus for the broader geopolitical changes that were to mark off the 1970s from the earlier postwar era. It will be recalled that, in late 1968, the Chinese had begun to signal both their perception of a receding United States threat in Asia and their interest in exploring a new relationship with Washington. The North Vietnamese, for their part, saw matters in an understandably narrower perspective. The military withdrawal program begun in mid-1969 in the name of the Nixon Doctrine and Vietnamization might have been seen by Chinese as confirmation of their perception, but Hanoi was determinedly intent on frustrating this program. What Hanoi feared was that the United States might successfully extricate itself from the grinding ground combat, thereby removing the domestic and international pressures on Washington that the North Vietnamese hoped to exploit, while leaving intact an anti-Communist regime in Saigon possessing a powerful military apparatus that could indefinitely postpone completion of the Communist revolution. Accordingly, the North Vietnamese were stubbornly insistent on an inseparable linkage of what they termed the two key points in a negotiated settlement, namely the military demand for complete United States withdrawal and the political demand for establishment of a coalition government to replace the intransigently anti-Communist one in Saigon. It was this linkage on which the Chinese and North Vietnamese increasingly diverged, even as it was this linkage that bedeviled the negotiations between Hanoi and Washington.

At this early stage, the Chinese made a point of castigating Nixon Administration policy and voicing strong support for the Vietnamese comrades, but it was the early signals pointing in new directions that were significant. Thus, at a banquet on 8 October for the high-level Viet Cong delegation present in Peking for PRC National Day, Chou En-lai[10] assailed the Nixon Administration for having stepped up its "counterrevolutionary dual tactics" of intensifying the arming of the South Vietnam military machine while "playing the deceptive trick" of withdrawing a small number of troops and proposing mutual withdrawals and internationally supervised elections. Significantly, however, Chou made an unusual reference to a Vietnam political settlement in saying the "sole correct road to a genuine settlement" was unconditional withdrawal of all U.S. troops. This remark came some three weeks after Washington's announcement (on 16 September) of the withdrawal of the second contingent of American troops. Peking's scattered references to U.S. troop withdrawal since Nixon made the first withdrawal announcement in June 1969 had not been made in the context of a political settlement. Chou's focusing on U.S. withdrawal presaged Peking's evolving approach to the Vietnam question, one that would come into conflict with Hanoi's linkage of the military and political demands. This divergence was already discernible at the time of DRV Premier Pham Van Dong's Peking visit later in the month. Citing a 10-point Viet Cong plan that was serving as the official negotiating position of the Vietnamese

Communists, Pham Van Dong stressed as the key points U.S. withdrawal and establishment of a Saigon coalition government. In the joint communiqué on the visit, the Vietnamese side, not the Chinese, called the 10-point plan "the correct basis" for a settlement, but the two sides together agreed only that "the sole correct road" to a settlement lay in an unconditional U.S. withdrawal.[11] In what would, in effect, provide feedback for the Chinese approach, Nixon soon thereafter (on 3 November) announced that a secret timetable for *complete* U.S. withdrawal had been established. Moreover, he released a midyear exchange of secret correspondence with Ho Chi Minh, and he disclosed that his representative at the Vietnam negotiations in Paris, Henry Cabot Lodge, had had 11 private meetings with the Communists apart from the regular, publicized sessions. If the Chinese did not see in all of this additional reason for positioning themselves for a post-Vietnam situation, Nixon's further revelation that United States officials had met with Soviet representatives to enlist Moscow's assistance in the Vietnam negotiations was bound to drive the Chinese into the bigger diplomatic arena to protect their own interests.

In speaking of Chinese interests, however, it is important to view these interests through the prism of Chinese leadership differences, and thus to appreciate the differential impact of various policy options on various factional interests. The initiatives taken by the Chinese in the wake of Czechoslovakia, and now being renewed following the decision to enter into border negotiations, had crucial implications for Peking leadership politics. As signaled in the September 1969 *Red Flag* article discussed above,[12] the question of whether to shift to a negotiationist track from the confrontationist politics of recent years bristled with implications for competing factional interests. Accordingly, a contextual analysis of that decisive shift must take into account the crucial internal developments that were integral elements of the context. Surely one of the most central elements was an institutional one, namely how the regime's institutional structure would be rebuilt in this post-Cultural Revolution period. Basic foreign policy questions could be addressed now that the border crisis had been brought under control; likewise, the outcome of the institutional crisis resulting from the shattering of the party apparatus during the Cultural Revolution posed fundamental questions regarding the shape of China's internal policy. It will be recalled that, in the autumn of 1968, during the period of various new initiatives in the wake of Czechoslovakia, a Central Committee Plenum had put on the agenda a new party congress, which was held in the month after the outbreak of border fighting. The same immobility that characterized Chinese foreign policy during the crisis period was reflected in domestic policy. In particular, the process of reconstruction of the party apparatus following the April congress had been painfully slow and indecisive. The institutional issue, so fraught with implications for high-level leadership politics, was now crying out for resolution.

At this stage, the cadre question—the question of who were to hold the reins of power in a reconstructed party following the tumultuous purges and bitter infighting of the Cultural Revolution—stood at the crux of the institutional issue. In an interesting pattern of timing, the month after the opening of Sino-Soviet border talks saw an upsurge of attention to this question. Two *People's Daily* editorials in November, the first editorials on a domestic topic since March and the eruption of border fighting, dealt with the cadre question, as did several articles in that month's issue of the ideological journal *Red Flag*. Expressive of the new sense of urgency being conveyed on the matter was a report from an Anhwei provincial meeting declaring that "the work of party consolidation must be given crash priority." An important breakthrough on the party rebuilding front was announced shortly thereafter, on 2 December, when Hunan reported the formation of a county party committee in Mao's native province, the first party committee to be formed at the county or higher level since the party congress. A year later, Hunan would also make the next trailblazing step by producing the country's first rebuilt provincial-level party committee.[13]

While the call to quicken the tempo of party rebuilding was resounding at this time, another and closely related development presaging major trends to come related to the military's role. There were now early warning signals that the PLA would be expected to devote itself increasingly to strictly military duties at the expense of the major political and administrative roles it had assumed following the collapse of the party apparatus during the Cultural Revolution. A significant signal was conveyed in an editorial on 9 December in the *Liberation Army Daily*[14] which indicated a growing attention to traditional military training and a reduction in the PLA's political role. In a notable gloss on the "four good" company movement, a campaign launched by Lin Piao in 1961 as part of his program to fashion the PLA into an exemplary institutional model, the editorial qualified the primacy accorded by Lin to the army's political and ideological work (the first of the "four goods") by warning that the three other aspects—work style, military training, and living arrangements—"must not be ignored" because "it is worthless to be good in one aspect." The editorial was concerned to adapt the "four good" principles to existing circumstances, making clear that a central concern in current circumstances was an urgent need for war preparedness. The institutional implications of this new direction—toward returning the army to its barracks and out of the political arena—were reflected at this time in renewed attention to the theme of party control within the PLA. Moreover, there were indications that the militia would be assigned a greater role in maintaining public order, a trend consistent with the withdrawal of the army to the barracks and an increased stress on military training. Thus, the issues were being drawn that early in the new decade would erupt into a monumental institutional and personal clash resulting in the demise of Mao's designated successor and a significant devolution of political authority from the military to civilian party institutions.

"Usher in the Great 1970s" was the heading of the joint editorial on New Year's Day, 1970,[15] heralding the decade in which the new trends being signaled at the end of the 1960s would come to the fore. The winds of change were given a rather heady expression in the editorial, which marked the first time since the onset of the Cultural Revolution that the New Year's editorial was devoted chiefly to international rather than to domestic topics. "Keeping the whole globe in view and looking ahead into the future," the editorial proclaimed, the Chinese were "full of excitement." Peking was, thus, buoyantly signaling to the world its intent to enter vigorously and innovatively into the international arena after the isolationist withdrawal of the Cultural Revolution era. Comporting with the premises underlying the new policy trends, the editorial portrayed the United States as having fallen from its postwar zenith, and it directed the main force of its attack on the Soviet Union, punctuated by Peking's first personal attack on Brezhnev since the opening of border talks—hardly a case of positive feedback for Brezhnev's irenic gesture toward "Comrade Chou En-lai." The editorial took care to reaffirm Peking's adamant position on territorial issues, saying Peking would seek to develop diplomatic relations with "all" countries on the basis of peaceful coexistence, "but on no account can we tolerate the invasion and occupation of our sacred territory by any imperialism or social-imperialism." This formulation covered both the Sino-Soviet border dispute and the Sino-American conflict over Taiwan, two issues that would provide touchstones for Peking's diplomatic progress in relations with the superpowers.

It was a notable characteristic of the arrival of the new decade that each of the three biggest powers was signaling its intent to enter what the Nixon Administration was calling an era of negotiation to replace an era of confrontation. As has been noted, this Nixon keynote had received explicit feedback in Gromyko's policy statement in July 1969, though it required a controlling of the Sino-Soviet border confrontation before the new era could begin to show tangible signs of existence. For the Soviet-U.S. relationship, the emerging theme of détente did not represent a radical departure from earlier phases, as one realizes by recalling such precedents as "the spirit of Camp David" during the closing days of the Eisenhower Administration, and the nuclear test ban treaty signed in 1963 in the wake of the Cuban missile crisis. Those precedents also served as a reminder that Peking, alarmed and embittered by what it perceived as a collaborative enterprise between the senior partner of its alliance system and the main enemy, had reacted by even more intransigently rejecting détente and withdrawing to a dual confrontationist policy that fundamentally transformed global alliance relations. Thus, for Peking now to enter an era of negotiation would mark a sharp shift in its policy line, the more so in that this had serious implications for Chinese domestic politics. It was the basic policy change by China that produced the geologic shift in the international political environment.

In contrast to the formal statement in November 1968 by which Peking had signaled its interest in exploring the possibilities of a new Sino-American relationship with the incoming Nixon Administration, the Chinese had recourse to a deceptively indirect mode of communication to signal their renewed interest a year later. This took place within a complex signaling process in which the Sino-Soviet border negotiations provided a central and reinforcing element. Shortly after the opening of those negotiations, Peking had used a Hong Kong channel, the Communist paper *Ta Kung Pao*, to convey its version of how things were proceeding in the negotiations and to counter an undesirable impression that things were going smoothly. In a dispatch dated 5 November, *Ta Kung Pao*[16] reported that "there is yet no sign of progress" in the talks. The report stressed the Chinese interest in a modus vivendi on the border by means of military disengagement, explaining that only in this way could steps be taken toward an overall settlement. The approach reflected in this leaked account of the talks comported with that set forth in the 7 October PRC statement announcing the agreement to open negotiations. In a further explanation of why this leak had been contrived, the dispatch said the Soviets should not "seek to attain other objectives under the cover of the Sino-Soviet negotiations." That cryptic reference reflected Peking's purpose in seeking to counter any impression fostered by Moscow that the negotiations were off to a promising start; having agreed to defuse the border crisis, the Chinese were concerned lest the Soviets, with the removal of the constraints imposed by that crisis, would be free to pursue diplomatic options elsewhere (for example, on strategic arms limitation and a post-Vietnam settlement) "under the cover" of the Sino-Soviet negotiations. In another case of interesting timing, the *Ta Kung Pao* dispatch was dated a day after Peking took its first public notice of the Soviet-U.S. announcement to open strategic arms limitation talks.

If Peking had followed past precedents, it would have pushed the Sino-Soviet negotiations to a breaking point and assumed a stiffly confrontationist posture toward the two superpowers. As events were to show, however, the Chinese were now positioning themselves for a notable flexible approach that would take them in a radically new direction. One element in this approach was to keep the Soviets engaged in border talks, not with any early prospect of a settlement but as a means of controlling Soviet pressure on China. The opening of talks and defusing of the crisis having been a precondition for Peking's new (or renewed) diplomatic initiatives, it was important to keep the negotiations alive (if barely) while opening up options on other fronts. This effort was reflected in Peking's handling of what it evidently perceived as at least a temptation on Moscow's part to put pressure on the Chinese to be either more amenable to progress in the negotiations or to have the talks suspended or downgraded. The occasion was the departure on 14 December of both Kuznetsov and his deputy on the Soviet negotiating team, V.A. Matrosov, for the ostensible purpose of attending the semi-annual session of the Supreme Soviet in Moscow. Both Peking and Moscow announced the Soviet negotiators' departure, but the announcements differed in meaningful ways. The Peking announcement pointedly cited the Soviet delegation as saying the negotiators would need to be absent for "about one week" and that therefore the talks had been "adjourned temporarily." The Chinese had thereby signaled their intent to resume the talks after only a brief interlude and had put the onus on Moscow to return its negotiators without undue delay. This was calculated to complicate any move by the Soviets to use a delay of the return of their negotiators as an instrument of pressure on the Chinese. In contrast to Peking's announcement, Moscow's version of the negotiators' departure gave no indication of the expected length of their absence and failed to note that the talks had simply been recessed.[17]

Peking made clever use of this context in signaling its interest in reopening the dialogue with the United States that had been aborted in February and further smothered by the border crisis atmosphere. In the very same transmission by NCNA's international service—Peking's primary channel for international political communication—which carried the announcement on the Soviet negotiators' departure, Peking juxtaposed a belated announcement that three days earlier, on 11 December, the Chinese chargé d'affaires and the American ambassador in Warsaw had met at the latter's request. The ambassador, Walter Stoessel, had first made contact with the Chinese envoy, Lei Yang, at a Yugoslav reception on 3 December and had extended an invitation to renew contact. Peking's terse two-sentence report on the 11 December meeting did not indicate any relation to the Warsaw ambassadorial talks, last held in January 1968, but the fact that the Chinese chose to juxtapose the announcements on the Soviet negotiators' departure and the Sino-American contact,[18] after delaying three days to report the latter, represented an early warning signal having profound implications for the approach to be taken by Peking in the new era that was dawning. Far from withdrawing into sullen isolation in reaction to Moscow's revitalized diplomatic activity in the international arena, Peking was now intent on ushering in the new decade by "keeping the whole globe in view and looking ahead into the future," as the New Year's Day editorial put it, a way of defining a broad geopolitical context for Peking's new initiatives.

Peking's exploratory signals toward the United States were receiving fuller feedback at this time than had been the case a year earlier. In November 1969 the United States had suspended naval patrols of the Taiwan Straits, a partial reversal of the crucial decision on 27 June 1950 (the date used by Peking to mark the beginning of United States "occupation" of Taiwan) to order Seventh Fleet interdiction of the Straits after the outbreak of the Korean War. A day after Peking's juxtaposed announcements of Kuznetsov's departure and the Stoessel-Lei meeting, Nixon announced that another 50,000 American troops would be withdrawn from Vietnam in April, thus following up his disclosure in November that a secret timetable for complete withdrawal had been established. And in another of the series of unilateral U.S. moves to relax restrictions on relations with China, Washington on 19 December lifted the $100 ceiling on Americans' purchase of mainland Chinese goods, thus furthering the liberalization of rules that began in the summer of 1969. In one of those silent signals that must be given due weight in any contextual analysis, Peking remained conspicuously silent on these moves by Washington, in contrast to its strident denunciation of efforts by the Johnson Administration to create an opening to Peking by relaxing restrictions on travel and trade relations.

The implications of Peking's semiotic juxtaposition of Sino-Soviet and Sino-U.S. announcements on 14 December began to become more explict early in the new year. On 9 January, Peking announced that Stoessel and Lei had met the day before to agree to hold the long-deferred 135th session of the Warsaw talks on the 20th. Washington at this time had another signal of its own for Peking, with State Department spokesman Robert McCloskey breaking a rigid taboo by referring publicly to the People's Republic of China by its own name,[19] a significant step toward according Peking the diplomatic legitimacy it was beginning to crave. Also on 9 January, the Hong Kong paper *Ta Kung Pao* again served as a channel for a Peking leak concerning the border talks. In an "extra" English-language edition,[20] the paper cited "well-informed circles concerned" for reaction to rumors from Moscow picked up by the Western press to the effect that there had been no progress in the border talks because the Chinese had devoted themselves to polemics against Soviet policies. The version conveyed by *Ta Kung Pao* was that, indeed, there had been no progress at the talks but this was due to Chinese insistence that the "understanding" reached between Chou and Kosygin in September on military disengagement should be honored. Only on this basis, according to the Chinese, could negotiations proceed without duress, but the Soviet Union had "refused to put any restraint" on its armed forces and had not assented to military disengagement in disputed areas. The notion of disputed areas was the sticking point, the Soviets being extremely loath to accept Peking's definition for fear that they would be acknowledging the validity of a territorial question and thereby risk a chain reaction of irredentist claims. Notwithstanding the disagreement on disengagement, however, there were no charges of border incidents, indicating that new unwritten rules along the border were being observed. The negotiations may have been deadlocked, and more than likely intentionally so on the part of the Chinese, but this was an element of a new equilibrium that sharply contrasted with the volatile instability of the border crisis before the Chou-Kosygin meeting. Perspectives could now be lengthened, permitting far-reaching negotiating lines to project into an evolving future.

For its part, Moscow was beginning to show a realization of the implications of the new developments. In the sharpest Soviet polemical thrust since the unilateral moratorium on direct polemics was imposed after the Kosygin-Chou meeting in September, a TASS report on 9 January[21] charged Peking with fostering a "military psychosis" in an effort to overcome internal strife and to purge elements conciliatory to the Soviets. TASS registered Soviet irritation over recent Chinese polemics beclouding the border negotiations, complaining that the Chinese editorial on New Year's Day contained "particularly vicious anti-Soviet attacks." It was significant that TASS attributed the anti-Soviet atmosphere to internal causes, and in seeking to probe Chinese leadership dissensions—a long-practiced but largely self-defeating Soviet approach— TASS went so far as to revive mention of Chinese Communists who allegedly favored good relations with Moscow. TASS singled out a 24 December *People's Daily* commentary as illustrative of the campaign "glorifying militarism and chauvinism in the spirit of Mao's thought." Interestingly reminiscent of the September *Red Flag* article discussed above,[22] the *People's Daily* article had assailed what it described as the revisionist view that there was a reasonable group in the enemy camp; and, as TASS took care to note, the article had approvingly cited Chiang Ch'ing—as had the

Red Flag article earlier. At this time, Moscow made its own use of juxtaposition to convey a message: *Izvestiya* juxtaposed the TASS account of Chinese militarization and the announcement (without adding comment) of the Sino-American agreement to reopen the Warsaw talks. Moscow was thus signaling its perception of the triangular nature of the developments now unfolding, but almost certainly Moscow misunderstood the Chinese internal dynamics impelling these developments, and consequently underestimated the depth of the change underway. It was not the radical confrontationist elements like Lin Piao and Chiang Ch'ing who were the proponents of this change; after all, the policies at home and abroad espoused by these elements represented lines of continuity with the dominant trends of Chinese policy in the second half of the 1960s. Rather it was the negotiationist approach represented by Chou En-lai that would fashion an instrument of anti-Soviet struggle immeasurably more effective than previous policy. In this respect, Brezhnev's unusual conciliatory gesture to Chou a week after the opening of the border talks proved to be grievously misdirected.

Awareness of the triangular nature of developments was expressed more explicitly in a 15 February *Pravda* article by S. Tikhvinskiy,[23] a Soviet Sinologist who was a member of the original delegation sent to Peking to open border negotiations in October. The article appeared a day after the anniversary of the Sino-Soviet treaty of alliance, which passed by without public notice by either side for the fourth successive year. The thrust of the article was a vigorous attack on Western speculation concerning a Soviet military strike against China, Tikhvinskiy taking particular exception to a book by Harrison Salisbury, *War Between Russia and China*, published in 1969 at the time of the border crisis. Tikhvinskiy interpreted works such as Salisbury's to be appeals to Washington to normalize relations with Peking and as also directed to anti-Soviet elements in Peking receptive to United States overtures. The *Pravda* article presented a historical review of Sino-Soviet relations to counter the picture of "age-old and irreconcilable" antagonism between the two giants on the Eurasian land mass, portraying, in contrast, an unceasing Soviet effort "to deliver China from the yoke of the imperialist powers." Using the attack on Salisbury's book as a peg for presenting Moscow's position on the border conflict, Tikhvinskiy accused Salisbury of falsifying the history of the border issue by repeating "the slanderous fabrications of Chinese nationalist propaganda on the supposed 'exploitation' of China by the Soviet Union, on certain 'territorial seizures' by the USSR in China, and on mythical Soviet claims to Chinese lands." The article was implicitly addressed to the "internationalist" elements in Peking in opposition to "nationalist" elements inclined to having dealings with the West against Moscow. This was the message in Tikhvinskiy's fervent protestations of Moscow's peaceful intentions and his rebuttal of those in the West attempting "to intimidate the Chinese leaders with the bugbear of a Soviet threat and to push the PRC into the arms of the United States."

The Tikhvinskiy article appeared at a time when the Sino-American dialogue seemed to be reopening with notable swiftness, in contrast to the deadlock that had immediately formed in the Sino-Soviet negotiations. The long-delayed 135th session of the Warsaw talks, held on 20 January 1970, was followed a month later by the 136th session, and at the later meeting the U.S. side sought to impart strong momentum to the exploration of a new relationship by proposing that a senior American offical go to Peking as a demonstration of Washington's seriousness about improving relations.[24] The evolution of a new relationship was still at a quite incipient and hardly irreversible stage, and the volatile Indochina situation would again boil up in new turbulence, but the logic of events was such that the bold breakthrough envisioned in that proposal would be achieved in a mere year and a half later. Henry Kissinger's startling trip to Peking in July 1971 produced an equally astonishing invitation to a much-reviled villain of old, Richard Nixon, to come to China as the representative of the erstwhile enemy of the people. Such was the force of the logic of events deriving from the premises established in the year and a half following the invasion of Czechoslovakia.

TWELVE

Transformation of Adversary Relations

The 20th anniversary of the Sino-Soviet treaty of alliance on 14 February 1970 would normally have been a major event in the Communist world. Not only are 20th anniversaries accorded major celebrations, but that treaty had special historical significance as the formal expression of Mao Tse-tung's decision in 1949, as the Chinese Communist armies swept toward final victory, to "lean to one side" in the bipolar postwar international system which by that time had firmly taken shape. In one of his only two trips outside China, Mao had gone to Moscow to negotiate with Stalin the terms of that alliance, in which Peking became a junior partner to Moscow in a vast and powerful "socialist camp" astride the Eurasian land mass. In his other foreign trip, in the autumn of 1957, a year after the Hungarian crisis, Mao returned to Moscow intent on reaffirming even more insistently the crucial role of the socialist camp in a bipolar system. Now, in 1970, the silence with which Peking and Moscow passed over the anniversary of the Sino-Soviet alliance for the fourth consecutive year offered mute testimony to the transformation of alliance relations that had taken place in the previous decade following the failure of expectations of the two partners in alliance. On the day after that 20th anniversary, *Pravda* expressed acute apprehension over the implications of the border crisis and of the new opening of Sino-American communication.[1] The border confrontation had made manifest how the

Soviet Union had been transformed from erstwhile ally into China's main enemy. Now the evolution of a new Sino-American relationship would further this process of adversary transformation.

Two developments in Indochina in the first two years of the new decade would pose a challenge to this process but, in their outcome, would actually serve to confirm it, as was dramatized by still a third development the following year. In the spring of 1970, the Indochina battlefield was substantially enlarged when Prince Sihanouk's reign in Cambodia was overthrown and that country was drawn deeply into the flames of war raging from Vietnam. As had happened a year earlier, the promising new Sino-American dialogue was interrupted, and Peking seemed to revert to a hard line against the United States that called into question the prospect of a new relationship. Next, in early 1971, a U.S.-supported drive into southern Laos by the South Vietnamese army elicited a sharp cry of alarm from Hanoi, which prompted the Chinese to warn the United States against going too far and to declare firm backing for their apprehensive comrades in Indochina. A complex signaling process proved effective in preserving the conditions for a new Sino-American relationship, and the way was soon open for the Kissinger trip and the invitation to President Nixon. Finally, in the spring of 1972, the Nixon visit to China having taken place, Washington undertook forceful measures against North Vietnam that included the mining of Haiphong harbor, an action the United States had theretofore avoided for fear of provoking a collision with the Soviets and driving Moscow and Peking back into alliance. Notwithstanding this action, which for years had been regarded as dangerously provocative, Moscow proceeded in the very same month, without even a cosmetic delay, to play host to Nixon for a summit meeting to match the Chinese. And thus was the geopolitical logic of events dramatized for all the world to witness.

The Cambodian convulsion began with the overthrow of Sihanouk on 18 March 1970 while he was out of the country. He went to Peking the next day and subsequently made it his base while acting as the titular leader of a government-in-exile and a national liberation front. Though the Prince originally declared his intent to "live in exile alternately in Moscow and Peking," the Chinese quickly embraced his cause, while the Soviets were discreetly distancing themselves and maintaining links with the new regime in Phnom Penh. Seizing the opportunity to use Cambodia as an instrument of Chinese influence in Indochina at the expense of the Soviet Union, Peking developed a theme of Asian Communist unity embracing the anti-U.S. forces in Indochina as well as North Korea and China. The Chinese arranged an Indochina summit meeting of solidarity around Sihanouk on 24–25 April at which Chou En-lai portrayed an Asian revolutionary struggle in which the anti-U.S. forces would "stand together, support, and assist each other and wage a common fight until the U.S. aggressors are completely driven out of Taiwan, South Korea, and the three Indochina countries." If the Chinese now had an opportunity to make capital out of Indochina developments, they were given further cause at this time when the United States and its South Vietnamese allies conducted a drive into Cambodia designed to clean out refuges and staging areas in that country used by the Vietnamese Communist forces. Peking reacted abusively and vociferously, calling Nixon "an extremely ferocious war criminal" whose "arrogant, unreasonable, Machiavellian and brazen features" had been revealed in his 30 April speech announcing the Cambodian "incursion."[2]

Not surprisingly in this context, the Chinese called off the session of the Warsaw talks scheduled for 20 May. Nevertheless, the Chinese took care to keep the Sino-American communication lines open, in significant contrast to their abrupt closing of the door at the time of the cancellation of the Warsaw session scheduled for February 1969. The announcement by Peking at that time had taken the occasion to denounce the "vicious features" of the new Nixon Administration and to accuse it of following its predecessors in "making itself the enemy" of the Chinese; that announcement said nothing about rescheduling the talks. The May 1970 announcement, on the other hand, avoided mentioning Nixon by name and did not refer to Sino-U.S. relations, while explaining that Peking deemed it "no longer suitable" to hold the meeting "as originally scheduled" in view of "the increasingly grave situation created by the U.S. Government, which has brazenly sent troops to invade Cambodia and expanded the war in Indochina." The announcement pointedly broached the question of rescheduling the meeting, saying this would be decided upon later through consultations.[3] Thus, Peking managed to register protest against the United States move into Cambodia, implicitly presenting a contrast with Moscow's uninterrupted participation in the strategic arms limitation talks with the United States, while at the same time signaling an intention to renew the Sino-American dialogue as circumstances permitted. None of this was, of course, lost on the Soviets, who made a point of juxtaposing two TASS dispatches, one datelined Peking and the other Washington, which simply quoted the Peking announcement, on the one hand, and, on the other, cited State Department spokesman Robert McCloskey as saying the Chinese had indicated that a new date would be discussed later and as having "implied that unofficial contacts with the Chinese side would be continued."

Instead of holding the Warsaw meeting on 20 May, Peking chose that date to issue a rare statement by Mao (the first since April 1968) in an effort to maximize the profits from its investment in Sihanouk and the Cambodian insurgency while the Kremlin hedged its bets and the Nixon Administration was engulfed in a flood of protest at home and abroad. Entitled "People of the World, Unite and Defeat the U.S. Aggressors and All Their Running Dogs,"[4] the statement introduced a new Maoist leitmotif in declaring that, though "the danger of a new world war still exists . . . revolution is the main trend in the world today." This was a gloss on the formula in the 9th Congress political report regarding a possible world war that said there were "but two possibilities"—either a war would give rise to revolution or revolution would prevent war. The revised version reflected a shift of focus from the Sino-Soviet confrontation to the Indochina situation and signaled Peking's attempt to capitalize on the turmoil arising from the Cambodian conflagration. The Chinese followed up five days later by signing a supplementary aid agreement with the North Vietnamese, the first such special agreement supplementing the annual aid accords and the first one in which Peking's announcement made a point of specifying the inclusion of military as well as economic aid. The new Sino-American dialogue was being drowned out by the clamorous battle cries of Indochina. On 20 June, a month after Mao's anti-U.S. statement had taken the place of the scheduled Warsaw meeting, Peking announced that its liaison official in Warsaw had informed his American counterpart that it would not be "suitable" at the present time to set a date for the next session; this would be discussed "at the proper time" by the liaison officials. The announcement pointed out vaguely that "both sides clearly understand the current situation."[5] In other words, Peking was still not severing the lines of communication but, rather, was signaling that the paroxysms in Indochina would need to be brought under control before the virulent adversary relationship between the PRC and the U.S. could undergo basic change. Though that was still some distance away, a signal in that direction appeared shortly after the United States withdrew from Cambodia in late June when Peking announced on 10 July the release of aged Bishop James Edward Walsh, an American who had been arrested for espionage in 1958. Peking coupled this announcement with the disclosure that "another U.S. imperialist spy," Hugh Francis Redmond, had committed suicide three months earlier.[6] The Chinese had thus cleared away some of the burden that had been encumbering any change in Sino-American relations.

Another area that needed to be set in order was the Chinese internal situation and the institutional reconstruction required by a rebuilding of the party. In essence, this meant a change—which would proceed from the painful to the convulsive—from the anomalous situation produced by the Cultural Revolution to the re-emergence of a party apparatus exercising institutional hegemony. This process required not only the rebuilding of the party but also the removal of power and prestige from those who had risen in the Cultural Revolution upheavals at the party's expense. Significant signals that would serve as bugle calls for the drawing of battle lines were transmitted at the time of the CCP's 49th anniversary in mid-1970. The 1 July joint editorial on the anniversary, entitled "Communist Party Members Should Be Advanced Elements of the Proletariat,"[7] looked forward to a year of accelerated party reconstruction in its concluding call to greet the 50th anniversary with "great achievements" in party-building. Appearing a few weeks before a crucial Central Committee Plenum that purged Ch'en Po-ta, the leading Maoist ideologue of the Cultural Revolution, who had risen to fourth position in the hierarchy, the editorial enjoined the party not to place in leadership position those with factionalist tendencies, a reference to Cultural Revolution zealots, and signaled the need for rehabilitation of experienced cadres who had fallen in the Cultural Revolution. In a pointed quotation from Mao that seemed directed at Ch'en, the editorial warned: "One must not always think himself in the right, as if he had all the truth on his side. One should not always think that only he is capable and everybody else is capable of nothing, as if the earth could not turn if he were not there."

The July issue of the ideological journal *Red Flag* also helped prepare the ground for rehabilitation of Cultural Revolution victims as part of the party reconstruction process. One article took note of controversy over the cadre question, especially the tendency of thinking that the Cultural Revolution had been a test and that those who had passed should now be in authority and others excluded.[8] Using a characteristic formula signaling a shift in line—the theory that another deviant tendency may be concealed in the main one and the charge that an errant line was being pursued in the guise of a proper one—the article criticized a tendency toward sectarianism and factionalism (leftist deviations) in the guise of the struggle against the idea associated with Liu Shao-ch'i that party members should be docile tools. The article inveighed against factionalism infecting some persons who, in dealing with party members having divergent views, resorted to criticizing and discrediting them for factional purposes. A crackdown against Cultural Revolution activists in the interest of party reconstruction was also reflected in another article in the July issue of *Red Flag*, which condemned those members of the masses "who had been seriously affected by the anarchist trend of thought" and "wanted to participate in every meeting of the party and inquire into the party's business." The article used a parable of the experience of a Shanghai plant's rebuilt party committee in dealing with representatives of the masses who resented losing their status and power to party organs that included cadres who had once fallen before the revolutionary masses' assault. Signaling that the party now meant business in recovering the reins of authority, the article stated bluntly that, while "there was a time when our party committee adopted a compromising attitude toward interference from the left," a "line of demarcation" must now be drawn between relying on the masses and succumbing to "tailism," which would nullify the party's leading role.[9]

In an unusual delay symptomatic of the troubles of this transitional period, Peking did not release the communiqué on the Central Committee Plenum held from 23 August to 6 September 1970 until three days after the plenum closed.[10] The communiqué introduced a benchmark of progress toward institutional reconstruction in recording the plenum's judgment that there was a "fervent desire" to convene the National People's Congress—a natural sequel to the April 1969 Party Congress—and proposed that necessary preparations be made so that the NPC could be convened at "an appropriate time." That the NPC is a rubber-stamp institution is precisely why it is significant in the signaling process, for to hold an NPC session means that important high-level decisions have been hammered out and are ready for general dissemination. The long failure to hold an NPC, despite repeated pledges, reflected the deep cleavages in the leadership and its failure to produce the necessary consensus, not the least on important leadership titles that an NPC would ratify. The plenum's purge of Ch'en Po-ta removed one major obstruction from the new trends that were encountering serious resistance from those having vested interests in the Cultural Revolution and its aftermath. The plenum's communiqué struck a keynote of the campaign against Ch'en in calling for study of Mao's "philosophical works" and opposition to "idealism and metaphysics," a reference to the leftist ideology regnant during the Cultural Revolution. This campaign proved to be a prelude, and served as a warning signal, of the convulsive struggle that would erupt in the Lin Piao affair a year later.

Also at this time, Peking identified Li Te-sheng as Director of the army's General Political Department, a post that had been vacant for three years. Li's appointment, together with other indications that the GPD was being rebuilt, signaled its reentry into the political process and, more significantly, the attempt to reconstitute party control over the military, an essential dimension of the Lin Piao affair. Still another warning signal was a Tsinghai broadcast on 27 November of an article by a political commissar of a PLA unit who was not a professional military man but a civilian and secretary of the unit's party committee. This, the orthodox pattern prevailing before the military establishment's rise to political power out of the chaos of the Cultural Revolution, proved to be a distinctive practice after the fall of Lin Piao and his top military associates.[11]

The highly significant trends in Chinese domestic politics emerging in the latter half of 1970 had their counterpart in foreign affairs. On the Vietnam issue, with its potential as a spoiler of new trends on the international scene, the United States on 7 October announced "a major new initiative for peace" that, in effect, promised an opening for Peking to pursue a more flexible and broadly based foreign policy. In expressing readiness to negotiate "an agreed timetable for complete withdrawals as part of an overall settlement," President Nixon said the United States was prepared to withdraw "all our forces" as part of a settlement based on "the principles I spelled out previously and the proposals I am making tonight." The United States initiative came after Kissinger had twice gone to Paris that September for secret meetings with North Vietnamese negotiator Xuan Thuy.[12] The significance in the initiative offered by Nixon lay in the calculated ambiguity of the formulation on troop withdrawal, in effect inviting the North Vietnamese and their patrons to make use of the ambiguity—which avoided explicitly demanding mutual withdrawals, an issue on which Hanoi was adamant—to open up the negotiating process. Hanoi waited a week before issuing a Foreign Ministry statement that resolved the ambiguity in an unpromising direction by interpreting the proviso that U.S. withdrawal would be based on principles previously spelled out as being a renewal of the demand for mutual troop withdrawal. The Soviets remained, in effect, noncommittal in criticizing the vagueness of the formulation while noting that the essence of Washington's past position had been mutual withdrawal. Peking's NCNA on 10 October, though saying Nixon's initiative was a "deceitful trick," noted specifically that he had not this time, as before, mentioned mutual withdrawals. On the 13th, a *People's Daily* "Commentator" article stressed Peking's line that the key to a settlement lay in an unconditional U.S. withdrawal. To the extent that the United States could demonstrate its readiness to withdraw without requiring mutuality, an area of divergence would open between Peking and Hanoi, the latter being committed to a linkage of its military demand (U.S. withdrawal) and its political demand (ouster of President Thieu). Indeed, it was to be this divergence, along with a parallel one between Moscow and Hanoi, *and* the evolving parallel relationships Washington developed with Peking and Moscow, that would apply intolerable pressure on this linkage and permit a Vietnam agreement.

If the Vietnam issue represented one major obstruction to new relationships on the international scene, the Taiwan question represented another seemingly intractable problem that had frozen Sino-American relations in deep hostility. On this question, real evidence of progress was evident within days of the new Nixon proposal on Vietnam. The PRC and Canada disclosed on 13 October that they had agreed to establish diplomatic relations based on a significant compromise formula for coping with the Taiwan issue. In the joint communiqué[13] the PRC reaffirmed that Taiwan "is an inalienable part" of its territory, but Canada was permitted to limit itself to saying it "takes note of this position" without endorsing it. Canada added that it recognized the Peking government as "the sole legal government of China," but this too made no commitment on the status of Taiwan. A *People's Daily* editorial on the agreement to establish relations used a different formulation, saying Peking was "the sole legal government representing the entire Chinese people."[14] The latter formulation was accepted by Equatorial Guinea in the joint communiqué a week later establishing that country's diplomatic ties with Peking, and appeared in subsequent agreements with countries also having less diplomatic interest or strength than a Western country in standing up to Peking. In the wake of the PRC-Canadian accord, Peking began signaling to the Japanese that they should get on the bandwagon, as in NCNA reports citing Japanese sources saying the news had aroused uneasiness within Japanese Premier Sato's circles and had strengthened the trend within the ruling Liberal Democratic Party toward establishing relations with Peking.

The significant breakthrough in Peking's relations with the Western world introduced by the Canadian recognition formula had a counterpart at this time in Chinese relations with the Communist world. This was expressed in relation to Yugoslavia, the erstwhile scapegoat of Peking's insistence on a bipolar international system but recently a beneficiary of the new trends appearing in the wake of the invasion of Czechoslovakia. On Yugoslav National Day, 27 November 1970, Peking introduced a formulation of the doctrine of peaceful coexistence that justified a notably flexible, differentiated line toward Communist countries. Speaking at a reception given by the Yugoslav Ambassador and attended by PRC Vice Chairman Tung Pi-wu and Vice Premier Li Hsien-nien, Vice Foreign Minister Ch'iao Kuan-hua declared that the five principles of peaceful coexistence underlay the improvement of Sino-Yugoslav relations and should apply to relations between "all countries, whether they have the same or different social systems."[15] This clearly diverged from the orthodox doctrine, to which Moscow subscribed, according to which peaceful coexistence should govern relations between countries having different social systems while proletarian (or socialist) internationalism should be the basis for relations between Communist countries. That doctrine was integral to Moscow's claim to a right to intervene in the affairs of other Communist countries, thus limiting the other countries' sovereignty in the higher interests of proletarian internationalism. It should be noted that the heterodox doctrine now being advanced by Peking had also been used by the Chinese in 1956 during a period of notably flexible policy toward both the West and the East European countries in Moscow's sphere.

The same October that saw Nixon's new Vietnam initiative and the Canadian breakthrough on the Taiwan question also happened to include the 20th anniversary of the Chinese entry into the Korean War, which had been a crucial event in forming the intense adversary relationship between the PRC and the United States. As a quinquennial anniversary, this one offered an occasion for the Chinese to make new contributions to the signaling process now underway, and they in fact took the occasion to do so. While marking the event with voluminous propaganda testifying to the past year's renewed warmth between Peking and Pyongyang, the Chinese at the same time reflected caution in defining their commitment to the North Korean cause, in contrast to Pyongyang's attempt to play on themes linking the two countries' vital interests. These divergent appraisals of matters of mutual security and vital interest reflected Peking's move into the big-power diplomatic arena, a move that would lead it to loosen its commitments to the smaller Communist powers like North Korea and North Vietnam. Moreover, Peking at this time began to refer to "peaceful unification" of Korea, an issue with sensitive implications for the Chinese because of the Taiwan question. Previously, Peking had failed to reciprocate Pyongyang's references to its program for peaceful unification; thus, in the extensive commentary surrounding the twentieth anniversary in June of the outbreak of the Korean War, when Pyongyang issued a lengthy memorandum on peaceful unification, the sole Chinese reference to the memorandum came in a speech by Chou En-lai in which he expressed support for the Koreans' "struggle for the reunification of their fatherland"—conspicuously omitting the reference to a peaceful process.

The most significant event in Peking's commemoration of this major anniversary was a rally on 24 October addressed by PLA Chief of Staff Huang Yung-sheng,[16] second only to Lin Piao among the military figures in the top leadership of the PRC. Huang gave an unusually—and thus meaningfully—explicit explanation for the Chinese intervention in 1950 having important implications for the present. He clearly distinguished between the period when the Chinese "supported and assisted" the North Koreans in "their fatherland liberation war" and the period of direct Chinese intervention. This intervention, he explained, took place after the United States had "flagrantly extended the flames of its aggressive war to the Yalu River in disregard of the repeated stern warnings of the Chinese people and gravely menaced the security of China." The several elements cited—flagrant extension of the fighting to the Chinese border, disregard of repeated warnings, and a grave threat to China's own security—defined a quite high threshold of provocation and threat to China itself before intervention would be triggered. In a later passage discussing the contemporary situation, Huang used the formula "support and assistance" to pledge Chinese backing in case of a new war in Korea; this was the formula used for the low-risk role the Chinese had been playing in the Vietnam conflict. Huang's definition of a high threshold of direct threat to Chinese security interests before intervention was warranted accorded with the general tenor of Peking's treatment of the anniversary, which occasioned a relatively relaxed portrayal of U.S. hostility and an almost complete absence of attention to the Taiwan issue. In contrast, the North Koreans used the occasion to draw parallels between their situation and that of the Chinese vis-à-vis a hostile United States, and to call attention to the American presence on Taiwan and continuing Sino-American tensions.

While Peking was itself being the source of significant signals of new trends, it was also at this time rewarded with hopeful signs that the world was relaxing its suspicions of Peking. On 20 November, the annual vote in the U.N. General Assembly on the China representation issue yielded an unprecedented majority

(51-49) in favor of seating the PRC, though this result was frustrated by the resolution defining the issue as an "important question" requiring a two-thirds majority. As an important factor in the new trends, the United States began adjusting its line on the U.N. representation question, thus making it easier for other countries to switch to a position favorable to the seating of Peking. U.S. delegate Christopher Phillips took an ambiguous position in a context in which ambiguity itself was a signal. He argued that there was no reason to expel the Chiang Kai-shek regime from the United Nations, inasmuch as it effectively governed 14 million people in Taiwan, but he declared to his fellow delegates that the United States was "as interested as any in this room to see the People's Republic of China play a constructive role among the family of nations." Though the American delegate emphasized that the United States firmly opposed the ouster of the ROC, the ambiguity of these signals would be resolved a year later by a sizable majority of U.N. member states in favor of seating the PRC at the expense of the Nationalist regime. Shortly after the narrow 1970 vote, President Nixon had said at a press conference that the U.S. had "no plans to change our policy with regard to the admission of Red China to the United Nations *at this time*," but he added that the U.S. would continue its initiatives toward Peking in seeking to open channels of communications and "eventually relations with Communist China."[17] These signals were even more forthcoming in Nixon's foreign policy report to Congress released on 25 February 1971, which repeatedly referred to the Peking regime by its official name, People's Republic of China, and said that there would be "no more important challenge" in the new decade than that of drawing the PRC into "a constructive relationship with the world community." The report reaffirmed the treaty commitment to Nationalist China, and it expressed the conviction that the differences between Peking and Taipei "must be resolved by peaceful means," but it said the United States relationship with Taiwan need not constitute an obstacle to "the movement toward normal relations" between Washington and Peking. Not surprisingly, given Peking's keen sensitivity to anything smacking of a two-Chinas solution that would legitimize the alienation of Taiwan indefinitely, NCNA's account of the Nixon report singled out the passages on the treaty commitment to the ROC and opposition to the latter's expulsion from the U.N. as showing that Nixon was still "playing his criminal 'two Chinas' plot." However, in a subtle acknowledgment that the studied references in the report to the PRC by its official name had been picked up as the signals that they were, NCNA quoted the observation that the 22-year-old hostility between the U.S. "and the People's Republic of China is another unresolved problem."[18]

One significant communications channel at this time was Edgar Snow, Mao's biographer from back in the Yenan cave days, who quoted Mao as saying in 1970 that Nixon would be welcome to visit China because "the problems between China and the U.S.A. would have to be solved with Nixon."[19] Before Snow departed from China, his hosts had made clear that he was speaking with authority. On Christmas Day (a subtle gesture of *bona fides*) 1970, NCNA reported that Mao had "recently" (meaning that it was some time ago but had been saved for publication now) met with Snow for "a cordial and friendly talk," and the *People's Daily* front-paged a photograph of Snow with Mao on the rostrum of T'ienanmen Square on National Day, 1 October. The messenger's credentials were well authenticated.

As the expansion of the Vietnam War into Cambodia in the spring of 1970 had posed a challenge to new dialogue opening between the United States and China, now in early 1971 an expansion of the war into southern Laos would do likewise. Here was a situation of complex communication patterns in which the signaling arising in an Indochina context had wider implications for the emerging Sino-American relationship. In February, the South Vietnamese army with U.S. support launched an operation (code-named Lam Son 719) in southern Laos aimed at crippling the North Vietnamese supply lines passing through that area. Peking reacted by issuing a Government Statement on 12 February that, for the first time in several years, directly linked China's security to military developments in Indochina. Claiming that United States "aggression against Laos is also a grave menace to China," the statement warned that "the Chinese people absolutely will not remain indifferent to it."[20] Two days later, a *People's Daily* editorial followed up by addressing itself to Washington's denial that the Laos operation posed a threat to China. "The new war venture of U.S. imperialism in Laos definitely poses a grave threat to China," the editorial insisted, adding a warning that the Chinese would "never allow U.S. imperialism to expand at will the war in Laos and the whole of Indochina."[21] Rallies beginning in Peking that day spread to other major cities. In addition, on 15 February, the Chinese signed another supplementary aid agreement with the North Vietnamese, as they had done in the previous May in the wake of the Cambodian operation. The Chinese were concerned over the credibility of their warning signals, taking particular exception to Nixon's denial that the action in Laos posed a threat to China as an attempt "to tie the hands of the Chinese people in giving support" to their friends in Indochina. In language reminiscent of Huang Yung-sheng's account the previous October of the conditions triggering Chinese intervention in Korea, an authoritative commentary claimed that the United States was "spreading the flames of war to the door of China" and lectured the President on the geopolitical realities of a common Sino-Laotian border of several hundred kilometers.[22] At this time, in fact, the Chinese hinted at an analogy with the Korean War, though taking care to avoid drawing the parallel on their own authority and to limit their commitment to the pledge of continuing rear-area "support and assistance"—the formula that, in Huang's account, covered the period before Chinese intervention in Korea.

During the first few weeks of Lam Son 719, the North Vietnamese expressed concern over the potential scope of military actions that might be directed against them, citing the Lam Son 719 operation along with a U.S. build-up by the demilitarized zone separating North and South Vietnam and in the Tonkin Gulf. Reacting sharply to Nixon's attempt to reassure Peking that Chinese security interests were not threatened, Hanoi became increasingly pointed in its effort to link its security interests with those of the Chinese and thereby to extract a stronger commitment from Peking. The Chinese warnings in February were designed to caution the United States to limit the operation to southern Laos. Likewise, Nixon in a press conference said that the Lam Son 719 operation presented no threat to China and that therefore the Chinese had no reason to "interpret this as a threat against them or any reason therefore to react to it." Nixon mentioned the Chinese presence in northern Laos, as if to suggest that each side could operate in that country without a confrontation. But the Chinese finally responded to Hanoi's concern by dispatching, in early March, a powerful delegation to Hanoi headed by Chou En-lai and including strong military representation. As they had done the previous month in the case of Laos, the Chinese, for the first time in years, directly coupled their security with that of the DRV. The Chinese purpose was twofold: to provide strong reassurance to the North Vietnamese, who had exhibited acute apprehension over U.S. intentions; and to warn the United States against attempting to retrieve a deteriorating situation in southern Laos by stronger moves directly against North Vietnam. To serve that purpose, the Chou delegation issued notably strong signals that the North Vietnamese eagerly picked up on and amplified. Thus, Chou pledged that the Chinese would "take all necessary measures, not flinching even from the greatest national sacrifices," to support the North Vietnamese—a warning of intervention that had disappeared from Chinese pledges of support in recent years. Speaking a day after Chou introduced this formulation, the DRV premier, Pham Van Dong, seized upon Chou's statement to deliver a warning to the United States to "remember the well-deserved lessons of [its] miscalculations all along the years"—a memory that would presumably include the Korean War.[23]

The joint communiqué on the Chou visit incorporated these strong warning signals, but it was couched in clearly conditional terms that indicated there were outer limits which the Chinese were warning the United States not to exceed. According to the communiqué, the two sides reached "completely identical views" on "how to deal with *possible* military adventures" by the United States, and the Chinese pledged all-out support, including the greatest national sacrifices, if the United States should "*go down the road of expanding*" the war in Indochina.[24] The direct linkage of Chinese and North Vietnamese security interests came in a passage saying the "new and extremely grave war escalation" by the United States "directly menaces the security of the DRV and at the same time the security of the PRC, thus creating a situation dangerous to peace in Asia and the world." This formulation, minus the reference to Chinese security, had been used by the North Vietnamese on the first day of the visit; Chinese willingness to link their own security interests with those of the Vietnamese was the central message the Chou mission served to deliver.

The Lam Son 719 operation and the surrounding political-military signaling proved to be a watershed in the Indochina conflict viewed as an international rather than a regional issue. It soon became clear that the credibility of Peking's commitments to Hanoi would not undergo a challenge by military developments in Indochina, and the Chinese found themselves free—almost exhilaratingly so—from the political encumbrances deriving from those commitments. At a Peking banquet on 16 March honoring the Chinese delegation's Hanoi visit, Chou En-lai described the current situation in Indochina as being "unprecedentedly fine," a new formula serving in effect to justify political moves by Peking unencumbered by commitments to allies in Indochina menaced by daring American military moves.[25] Three days later, at a banquet marking the anniversary of Sihanouk's arrival in Peking following his overthrow, Chou declared that the situation in Indochina after Lam Son 719 had "never been so favorable as it is today."[26] Unlike Chou, both the North Vietnamese ambassador, at the 16 March banquet, and Sihanouk, three days later, made a point of recalling Chou's pledge in Hanoi that the Chinese would not flinch from even the greatest sacrifices to support their Vietnamese allies. Those allies were soon to learn how divergent their perceptions and expectations were from those of the Chinese.

Peking moved quickly to capitalize on the "unprecedentedly

fine" situation. The first significant move was an exercise in "people's diplomacy," which began with an invitation to an American table tennis delegation to visit China. The invitation, delivered on 7 April to the Americans while they were in Nagoya, Japan, for the world championships, came less than a month after Washington had announced the removal of restrictions on American travel to the PRC. Now the time had ripened for a move by Peking reciprocating the series of unilateral United States initiatives since mid-1969 to relax restrictions on relations. The political signal contained in Peking's invitation was conveyed by Chou when he told the Americans on 14 April that their visit had "opened the doors to friendly contacts between the people" of the two countries and expressed the belief that "such friendly contacts will be favored and supported by the majority of the two peoples."[27] Of equal significance was the absence of any polemical intrusions in the course of the visit, apart from a discordant note sounded in a live Shanghai broadcast of matches there on 15 April when the announcer made an invidious distinction between the American people and their government. NCNA's account of the Shanghai matches, however, did not refer to the U.S. Government. Similarly, at a banquet given by Sihanouk on 13 April, the Prince denounced President Nixon as "our principal enemy," but Chou in his reply made no mention of Nixon while predictably expressing support for the Cambodian insurgents' struggle against U.S. "imperialist aggression.[28] For its part, the Nixon Administration was helping confirm Peking's perception of a sharply reduced threat in Vietnam by announcing on 7 April that the rate of troop withdrawals would increase in the period from May to December, bringing the total number withdrawn to more than two-thirds the number there when Nixon took office.

While the Vietnam War was becoming less and less a factor impeding the evolution of a new Sino-American relationship, the Taiwan question remained as the "crucial issue" in Sino-U.S. relations, according to Peking's definition. There were two essential elements in this issue: the juridical status of Taiwan, and the U.S. military presence there. Peking had deftly capitalized on the diplomatic flexibility afforded by the Canadian "take note" formula for circumventing the Taiwan issue, but, at the same time, it expressed acute concern lest the international community regard this relaxation of Peking's demands on the Taiwan issue as legitimizing a two-Chinas situation. As for the other aspect, the U.S. military presence symbolized what Peking called "occupation" of its territory; this related to the impasse between Washington and Peking over the former's insistence that the Taiwan issue be settled peacefully, and Peking's equally adamant insistence that this was an internal matter brooking no interference from outside powers. In an authoritative statement on the issue delivered on 27 June 1970 (the 20th anniversary of U.S. "occupation" of Taiwan), a high-ranking Chinese leader heading a delegation in North Korea to commemorate the outbreak of the Korean War had declared that "relaxation of Sino-U.S. relations is, of course, out of the question" so long as the United States refused to withdraw its armed forces from Taiwan.[29] Thus, a real breakthrough toward a new Sino-American relationship required an accommodation on an issue that, as an enduring legacy of the Korean War, had produced two decades of hostile alienation between two formerly friendly nations.

As part of Washington's effort to prepare the ground for handling this thorny issue, a Presidential commission headed by Nixon's 1960 running mate, Henry Cabot Lodge, submitted a report on 26 April 1971 that, in effect, endorsed a two-Chinas policy for dealing with the China representation issue in the United Nations. The new fluidity evident in Washington's China policy was especially reflected three days later when President Nixon, asked at a press conference about the Lodge report and the Administration's China policy, stressed that speculation in "this very sensitive area . . . might destroy, or seriously imperil, what I think is the significant progress" that had been achieved. But while understandably seeking to dampen speculation that would complicate a sensitive signaling and negotiating process, Nixon gratuitously volunteered —with the lame explanation that he knew "this question may come up if I don't answer it now"—the revealing statement that

"I hope and, as a matter of fact, I expect to visit mainland China sometime in some capacity."[30] This was an exceptional signal to Peking that Mao's message conveyed by Edgar Snow had been received and taken seriously, or at least hopefully.

The difficulties posed by the Taiwan issue were uncomfortably evident at this time in the rough day experienced by a State Department spokesman on 28 April when he was questioned closely by the press on the administration's China policy, an experience that may have prompted Nixon's plea the next day against speculation. The spokesman, Charles Bray, actually gave two press briefings that day, before and after noon, the second session having been held after he himself had been briefed by China policy advisers including a deputy assistant secretary of state. The whole exercise provoked a sharp retort from Peking, as reflected in the title of a *People's Daily* commentary, "Fresh Evidence of the U.S. Government's Hostility Toward the Chinese People."[31] Peking took particular exception to Bray's clarifying remark that the Cairo and Potsdam declarations during World War II, which said Taiwan should be returned to China from Japan, represented a statement of purpose that had never been formally implemented. The Peking commentary branded this as "barefaced lying," arguing that "the then Chinese Government" had assumed control of Taiwan in 1945 and that thus China had resumed sovereignty over Taiwan from that time on. In addition to his point about sovereignty over Taiwan remaining an unsettled question subject to future international resolution, Bray had also said this question should be settled by the two rival Chinese regimes. The latter idea was not examined on its own merits in Peking's commentary but was cited only in conjunction with the other point as indicating the "self-contradictory and nonsensical" character of Bray's remarks. As usual, Peking also noted that this represented interference in China's internal affairs and that "when and how the Chinese people liberate Taiwan is entirely China's internal affair." Peking also took the occasion to note how far apart it remained from Washington in coming to an accommodation on this issue. While acknowledging that there had been "a new development in the friendship between the American and Chinese people" since the ping-pong diplomacy, Peking observed scornfully that the Nixon Administration had "hastily made various gestures . . . as if to improve relations," and that Bray's briefing and other evidence demonstrated that Washington's professed intention to establish normal relations was "all humbug." Peking was signaling that Washington would need to do better on the Taiwan question. Chinese wariness was also reflected in a remark by Chou En-lai on 8 June during the Peking visit of Romanian leader Ceausescu, a communication link between Washington and Peking. Chou cautioned that "we must always keep a clear head about the imperialists, neither fearing their bluster, nor easily believing their 'nice words'." The Romanian mediatory role was reflected in the joint communiqué in a passage, attributed to the Romanian side only, that registered satisfaction over the "recent increasingly manifest tendency" of "certain capitalist countries" to normalize relations with the PRC.[32]

The sensitive nature of these trends toward a new relationship between long-hostile adversaries related to other considerations besides the signaling between the parties themselves. To be sure, that in itself was a very delicate process, each side having to provide enough positive feedback for the other's probing signals to maintain the evolving process, while at the same time reserving its positions and preserving its diplomatic and strategic assets in case the process was short-circuited. Still another dimension, however, was the internal one: each side required a mandate enabling it to take the initiatives, and provide the positive feedback, that would sustain the evolution. It has often been remarked that Richard Nixon was ideally suited, in view of his hard-line anti-Communist background, to head an administration that could take the initiatives required to transform the frozen adversary relationship with China; his flanks were protected where a President without such a conservative background would be vulnerable. In the case of the Nixon Administration, then, the delicacy of the situation derived mainly from the signaling interaction with Peking, on the one hand, and the need to reassure clients such as the Thieu regime in South Vietnam, the Chiang Kai-shek regime in Taiwan, and the Republic of Korea that this process would not endanger their positions. For the Peking leadership, however, the internal dimension remained of decisive importance, as one could readily appreciate upon recalling how the fragile mandate underlying the Chouist initiatives in late-1968 collapsed under the strains of the following year.[33] The plenum in August-September 1970, particularly the purge of Ch'en Po-ta, and the trends emerging in its wake emitted warning signals of instability in the Chinese leadership, an instability that might call into question the mandate for the moves on the international scene required for a new relationship with an old adversary. These trends were to continue to gather momentum, however, until their ascendance posed such a threat to important forces in the leadership that the convulsive Lin Piao affair erupted the next year.

An important milestone marking the progress of these trends was the establishment of the first provincial-level party body to be rebuilt since the dismantling of the party administrative apparatus during the Cultural Revolution. The first reconstructed provincial party committee (announced by Peking on 14 December 1970) was Hunan's, Mao's home province, and the first to report a rebuilt county-level party body the previous year. Later that month (31 December) Peking announced that three additional provinces had established rebuilt party committees on 26 December (Mao's birthday, though, according to custom, this was not explicitly noted). Against the background of these significant organizational developments, the 1971 New Year's Day joint editorial was notable for omitting a significant role for any institution except the party; there was no mention of the PLA's civilian role, a role that had been the hallmark of its ascendance to political power during the Cultural Revolution. At the same time, signaling the intent to maintain the momentum of party rebuilding (which had been conspicuously lethargic in the months immediately after the 9th Congress in April 1969), the editorial termed it "necessary" to "successfully convene the local party congresses at various levels," and to strengthen the party's democratic centralism and centralized leadership. It also indicated that a National People's Congress would be convened in that year.[34] The importance of democratic centralism, the controlling principle of orthodox Leninist party organization, was stressed in an article in that year's first issue of *Red Flag* by the trailblazing Hunan party committee, whose message was clearly intended to be taken seriously as expressing the trend of the times. This article, "Further Strengthen the Party's Democratic Centralism,"[35] criticized "some comrades" for regarding democratic centralism as opposed to unity of party committees, and it made clear a concern over abuses of power by military representatives heading party bodies who were arrogating decision-making to themselves and ignoring other elements in the "three-in-one combination" of old, middle-aged, and young cadres constituting the new bodies. The article also demanded that arbitrariness by other institutional sectors (meaning mainly the PLA) be overcome by the party's exercise of unified leadership. These themes could not fail to be perceived as danger signals by those military figures who had ascended to the top of the Chinese hierarchy on the ruins of the party during the Cultural Revolution and whose unprecedentedly extensive representation at the highest levels of the leadership had been confirmed at the 9th Congress.

A *People's Daily* editorial on 30 January,[36] coinciding with the announcement of the formation of the eighth new provincial party committee within eight weeks, offered Peking's first editorial comment on this surge of party rebuilding and marked the first editorial on a domestic topic in three months. Hailing the formation of the new party committees as significant advances in "strengthening the centralized leadership of the party," the editorial stressed the "great importance" of a good work style and warned against arrogance and rashness, qualities infecting the military's exercise of the political authority befalling it during the Cultural Revolution. Tensions arising from this process were reflected in the more explicit content of provincial media, as in a Yunnan broadcast (27 January) charging that some PLA personnel had "overcriticized the shortcomings of the civilian cadres," and a Hupeh broadcast (26 January) chiding PLA cadres in political posts for being "arrogant and unwilling to learn from the local cadres." On a more authoritative level, the target of the campaign now in full swing against arrogance among cadres was signaled in a *Red Flag* article's criticism of "some leading cadres" who were "obsessed with the idea that they had won victories—military victories during the war and new victories during the great Cultural Revolution."[37] Similarly, another article in the same issue, this one attributed, significantly, to the party committee of a military division, made it sharply clear that the problem of arrogance among military cadres was a serious one; one individual was criticized for his arrogant behavior as a leader of a support-the-left team in the Cultural Revolution, which had been a decisive instrument of military intervention in politics. The deleterious effect of one-man rule on democratic centralism was criticized, and the article took to task military figures who had a long background in revolutionary warfare and had "also passed the test in administrative work" during the Cultural Revolution but had become arrogant, complacent, and intoxicated with power.[38]

This campaign on organizational issues of democratic centralism versus arrogance and one-man rule was now to have an esoteric, and therefore more ominous, counterpart on the ideological level in the form of a campaign against "idealist apriorism," defined as a rival of dialectical materialist epistemology. Viewed as part of the signaling process, particularly as a source of warning signals, this exercise in esoteric communication was deadly serious business. The themes of the campaign were enunciated in an article whose authority for this purpose was attested by its publication not only in the theoretical journal *Red Flag* but also on the front page of *People's Daily* and *Liberation Army Daily*; also significantly, it was attributed to one of the rebuilt provincial party committees. Anticipating the line of attack on Lin Piao that developed in the immediate aftermath of his downfall, the article castigated views that were "extremely leftist" or "right in form but left in essence." In elaborating a theory of knowledge as an endless process of "practice, knowledge, again practice, and again knowledge," the article stressed the need for taking the initiative to readjust subjective thinking to make it suitable "to the new situation when the objective process has already progressed and changed from one stage of development to another." It inveighed against the "bourgeois work style and ideas of arrogance, complacency and getting into a rut found among some of our comrades" who considered themselves always in the right after having been so at one time or another.[39] This warning signal was conveyed on the most authoritative level in the 1971 May Day joint editorial in *People's Daily, Red Flag,* and *Liberation Army Daily,*[40] "urgently demanding" that party officials, "senior cadres in particular," strive to raise their theoretical understanding of Marxism-Leninism. Most notably, the editorial called for "criticism of revisionism and rectification of style," and for criticism of idealist apriorism and of other heresies ascribed to the disgraced Liu Shao-ch'i "and other sham Marxist political swindlers." This was not an arcane dispute over Marxist theory nor beating a dead horse; a major political crisis and power struggle was in the offing.

In addition to providing a backdrop for the political and ideological signaling that increasingly pointed to an impending crisis, the process of party reconstruction itself offered evidence of underlying troubles. As has been noted, this was evident in the painfully slow progress of party rebuilding after the 9th Congress, but it was also to be found after that rebuilding process had gathered strong momentum. In the first seventeen provincial-level party committees established, the chairman of the province's revolutionary committee, the heavily military-dominated organ which arose during the Cultural Revolution to replace the party apparatus as the administrative structure on the provincial and local levels, became the first secretary of the party committee. As of late May, however, in five of the last six provincial party organs formed, the revolutionary committee chairman failed even to be named to the secretariat. This was the period, it should be noted, when the campaign against idealist apriorism was developing and "sham Marxist political swindlers" were being attacked. This was also a period in which noteworthy rehabilitations of officials who fell victim to the Cultural Revolution's assault on the party were taking place; examples were the former Kwangtung and Hunan party chiefs, who reappeared elsewhere as secretaries of rebuilt provincial party committees. The marked deviation from the pattern that had obtained without exception for the first seventeen reconstructed party committees was followed by a period of no less than three months when there were no new provincial committees formed. The process of reconstituting the country's power organs was producing grave conflict. The diminished authority of the army, a development that in turn posed a threat to the status and power of those high leaders who had risen with the PLA's ascent to political authority, was clearly signaled on the PLA's anniversary, 1 August, when, for the first time in four years, the joint editorial[41] cited Mao's dictum that "the party commands the gun, and the gun must never be allowed to command the party"—an apt and highly significant reminder of a tenet that had received scant attention during an era in which the military had supplanted the party in the political and administrative apparatus of the country. Moreover, for the first time since the Cultural Revolution began, the editorial contained no quotation from Lin that all should heed, not even his injunction to study Mao's thought. Lest all of this was not sufficient to signal the direction of events, the editorial stressed that the army was "under the party's absolute leadership, going where the party directs." By late August, when the final four provincial party committees were formed at last, the Lin Piao affair was only days away from its sensational climax.

While this internal convulsion was gathering momentum, China's new differentiated foreign policy found its most dramatic expression in the secretly arranged visit by Henry Kissinger in July and the invitation extended to President Nixon. Those bare facts alone dramatized the depth of the changes taking place in the alliance and adversary relations shaping the international landscape, but Peking at the same time signaled that these changes derived from a fundamental conceptual rethinking of the international system and China's place in it. It was important that Peking invoked the highest authority, canonical Maoist scripture, to certify this significant shift from time-honored policies and perceptions. This was important in order to convey Mao's mandate for the shift by demonstrating its underlying continuity with the fundamentals of Maoist strategy. It was important to convey that mandate in order to foreclose a challenge from powerful forces whose interests would be adversely affected. The original impetus for the shift derived from the repercussions of the invasion of Czechoslovakia, but, as has been seen, the policy initiatives developing from that impetus had enjoyed only a fragile mandate that came undone under the stresses and pressures of 1969. Now, with the Lin affair about to result in a strengthening of the hand of Chou and others associated with these initiatives, the fruition was soon in coming.

A cue for invoking Mao's mandate for these initiatives was contained in the 1 August Army Day joint editorial,[42] which hailed the "great victories" being achieved by "Mao's revolutionary diplomatic line," an unusual formulation that might be termed a line of negotiation as the continuation of revolution by other means. That editorial was reprinted in the year's ninth issue of *Red Flag*, the first issue of the theoretical journal to be prepared since the Kissinger visit. Most notably, the same *Red Flag* issue contained a highly significant article calling for study of Mao's canonical 1940 work "On Policy" as guidance, at a time when the Chinese again found themselves "in the thick of an extremely complex situation."[43] The sharp policy changes culminating in the Nixon invitation certainly cried out for guidance and explication, but they were also highly sensitive matters ideally suited for just this sort of instruction by exegesis. Apart from the article's appearance in the authoritative journal, its importance was also underscored by its having been broadcast by Peking radio and reprinted in *People's Daily* on 17 August. Moreover, two weeks later NCNA disseminated it internationally, the time lag between domestic and foreign dissemination perhaps reflecting the sensitivity of the matter. Whatever the particular circumstances, an important message was being conveyed.

The essence of the message was a reordering of Peking's adversary relations by distinguishing between "the principal enemy" and "the enemy of secondary importance," a differentiation designed to isolate the principal enemy by capitalizing on "contradictions" among adversaries and making use of "temporary allies or indirect allies." The context of the original article, "On Policy," was the war against Japan and the need for a united front against that principal enemy of the time. Updated to the current context, this would mean seeking accommodations with the United States and its allies in an effort, as *Red Flag* defined the goal, to "force our principal enemy into a narrow and isolated position." In a crucial passage, recalling the period of Mao's 1940 work, *Red Flag* observed that the Chinese Communists had "made a distinction between Japanese imperialism then committing aggression against China and the imperialist powers which were not doing so at that time; and between the imperialist powers which adopted different policies under different circumstances at different times." The first distinction, justifying a concentration on the Soviet Union as the current principal enemy committing aggression along China's border, was lifted directly from "On Policy," but the second distinction telescoped a passage in the 1940 work that includes explicit references to China's relations with the United States. Heeding *Red Flag's* advice to turn to "On Policy" for current guidance, the reader finds that it distinguishes between U.S. policy in an earlier phase in the 1930s—when the United States "followed a Munich policy in the Far East"—and its abandonment of that policy by 1940 "in favor of China's resistance." This aspect was too sensitive to be spelled out in the current context, but its implications were clear. Helping to make the message clear, *Red Flag* stressed that the view that "all enemies are of one cut and are a monolithic bloc does not conform to objective reality." On the contrary, it was of utmost importance to "be good at seizing opportunities" to capitalize on the "rifts and contradictions in the enemy camp and turn them against our present main enemy." Hence the invitation to President Nixon in the wake of Czechoslovakia and the Sino-Soviet border crisis.

The *Red Flag* article's reference to a Far East Munich in the context of the Anti-Japanese War is reminiscent of a similar reference, for quite different purposes, in Lin Piao's September 1965 tract on people's war, which for its time had also served as a highly significant signal of Peking's policy direction at an important juncture. In the analogy implied in Lin's tract, Japan's role in the Sino-Japanese War was being played by the United States in 1965, and the U.S. role by the Soviet Union in 1965 (as the Vietnam War was reaching serious international proportions). Lin charged that the United States during the Anti-Japanese War had plotted a Far East Munich and had offered to aid China as part of an effort to turn China into a colony; this charge prefigured Peking's line that the Soviet offer of united action on Vietnam was intended to reduce China to a protectorate. By the time of the 1971 *Red Flag* article, the Soviet Union had become the analogue of Japan during the Sino-Japanese War as the principal enemy, and the United States was being credited with having gone beyond its "Far East Munich" policy to one "in favor of China's resistance" to Soviet hegemony. This was the essential basis of the new Sino-U.S. relationship soon to be enshrined in the Shanghai communiqué on the Nixon visit.

There was another notable difference between policy pronouncements in 1965 and the *Red Flag* explication of policy in 1971 that reflected the fundamental change in direction being signaled. A watershed policy statement in November 1965[44] foreclosing any accommodation with Moscow in behalf of the North Vietnamese had, like the *Red Flag* article, sought to give policy guidance for a "complex situation." But, where in 1965 the Chinese were instructed that they "must never abandon or slur over principles" by accepting even a limited accommodation with the Soviets, the basic thrust of the *Red Flag* article in 1971 was that "great flexibility" was required in order to isolate the main enemy. Interestingly, in light of the Lin Piao affair soon to explode, the *Red Flag* article delivered a sharp attack on those who formulate tactics according to revolutionary sentiment rather than using the flexible tactics prescribed in the article (the better to "adhere to firm revolutionary principles," as the article put it dialectically). The article's polemical insistence against substituting feeling for policy suggested that Lin may have appealed to ideological and moral sentiments in opposition to Peking's opening its door to the United States. But, whatever strength of opposition he may have previously mustered to undercut these new policy directions, Lin and his power base were now doomed to ignominious collapse.

The conceptual foundations had been laid for the transformation of the Sino-U.S. adversary relationship, to be expressed in the formulations contained in the Shanghai communiqué released at the conclusion of Nixon's visit on 28 February 1972.[45] In addition to positing the goal of normalization of relations between the U.S. and the PRC, that communiqué formulated a pair of key elements in this evolving new relationship: neither side "should seek hegemony in the Asia-Pacific region and each is opposed to efforts by any other country or group of countries to establish such hegemony"; and both sides "are of the view that it would be against the interest of the peoples of the world for any major country to collude with another against other countries, or for countries to divide up the world into spheres of interest." These basic elements reflected crucial changes in Peking's perceptions, or, more precisely, the perceptions of leaders now in the ascendant in Peking and enjoying Mao's mandate. One essential perception was that the United States was now no longer playing a hegemonial role in East Asia but was making adjustments that would, in effect, accommodate Peking's interests to a more or less degree. Another such essential perception was that there was no longer an underlying Soviet-U.S. congruence of interest in the workings of an international system that had served to subordinate China's own vital interests. In effect the parties to the Shanghai communiqué forswore, on the one side, a continuation of the Soviet-U.S. "collaboration" against China that Peking had so insistently attributed to the superpowers in the 1960s, and, on the other side, a return to the Sino-Soviet alliance that Americans had once perceived as the main threat to international security in the bipolar postwar environment. A perceived forfeiture on these counts would emasculate the Shanghai undertakings.

The Taiwan issue occupied a major part of the Shanghai communiqué, and the compromise reached was essential to the new relationship, but the role played by this issue must be analyzed on two levels. First, on the level of complete normalization of relations as a goal posited in the communiqué, the Taiwan question must, of course, be regarded as a central factor in bilateral relations. However, this was not the level on which the forces had been operating to transform the Sino-U.S. adversary relationship into one in which the goal of bilateral normalization could be envisaged. Those forces derived mainly, on the one side, from Washington's concern over the United States position as a result of the painful Vietnam experience, and, on the other side, from Chinese perceptions of Soviet hegemonial proclivities in the wake of Czechoslovakia and in the context of the Sino-Soviet border confrontation. In terms of this level the compromise on the Taiwan issue should be viewed as an enabling condition which served other, more direct and immediate purposes of the two sides.

From Peking's standpoint, the Shanghai communiqué's treatment of the Taiwan question represented a logical extension of the new flexibility signaled by the Canadian formula of October 1970. The other party was being allowed to reserve its position in order that the two sides could get on with the business of developing relations and broadening Peking's leverage in the international community. The United States side, however, went further than Canada and the other NATO allies following its lead by going on record explicitly as not challenging the position—carefully attributed to "all Chinese on either side of the Taiwan Strait"—that "there is but one China and that Taiwan is a part of China." As Washington knew, this was not a factual statement, for there was a well-known Taiwan independence movement of expatriate Chinese from Taiwan who in fact challenged precisely this position. The operative meaning of the statement was that the United States was committing itself to refrain from promoting or underwriting a formula that would juridically alienate Taiwan from the mainland, an option that had been proposed with increasing currency in recent years as a means of resolving the anachronistic impasse of two rival regimes claiming to rule over the same nation more than two decades after the civil war ended on the mainland. The United States had thus abandoned the position of a year earlier—which Peking had found so unacceptable—that the status of Taiwan remained undetermined.

The United States did, however, reserve its position on what it had long regarded as the crux of its concern in the Taiwan question, opposition to a forcible takeover. In counterpoint to Peking's reiteration of its own longstanding position that "liberation" of Taiwan was China's internal affair in which no other country had a right to interfere, the United States reaffirmed its "interest in a peaceful settlement" of the Taiwan issue by the Chinese themselves. It was with "this prospect in mind" that the United States committed itself in the Shanghai communiqué to a total withdrawal of its forces from Taiwan as an "ultimate objective." Thus, a hard kernel of disagreement remained unresolved and subject to further negotiation and evolution toward full normalization of relations, the United States having *conditioned* its full compliance with Peking's demand for military withdrawal on a peaceful settlement of the Taiwan issue, and Peking having adhered to the position that its sovereignty over Taiwan was *unconditional*. Nonetheless, the crucial bilateral issue had, in a few months, been brought into manageable control, thus enabling the two countries to transform their relationship for purposes that far transcended the bilateral framework.

Nowhere was this development more a matter of concern than in the Kremlin, which could not help being apprehensive over the enhanced leverage given to the Soviet Union's major rivals from the transformation of their adversary relations. In an interim assessment of the Sino-U.S. summit delivered a few weeks later, two months before Moscow was to host Nixon at a parallel summit, Brezhnev deferred judgment until concrete results were observable, results he suggested might be expected in the very near future. However, he did observe warily at this time that some statements made during Nixon's China visit indicated that its purpose went beyond a bilateral framework, which could hardly have been surprising to the Soviet leaders. Nonetheless, Brezhnev's basic message was that Moscow was itself open to exploring the negotiating path with both China and the United States, and he struck a forthcoming note in looking toward Nixon's visit as an opportunity to establish "mutually advantageous cooperation" which would not, he took care to add, be at the expense of third parties. Moreover, he also made an ideological concession to the Chinese by acceding to their heterodox line of placing Sino-Soviet relations on the basis of peaceful coexistence, the basis adopted in the Shanghai communiqué for Sino-American relations, and he disclosed that the Soviets were offering the Chinese "specific and constructive" proposals on non-aggression and a border settlement, a disclosure that coincided with the return to Peking of Soviet border negotiator Ilichev after a three-month absence. In short, Moscow was signaling its acknowledgment of the changes in the international landscape and its readiness to explore a Soviet role in the emerging new relationships.[46]

In the event, circumstances themselves caused Moscow to give that signal a rather persuasive form. During the period between the two summits, the Indochina War flared still another time, the third year in a row in which that legacy of the turmoil of the 1960s was to pose a challenge to the new trends emerging on the international scene. But where the Cambodia and southern Laos operations in the two previous years tested these trends in the Sino-American context, the spring of 1972 saw a challenge to Soviet-U.S. relations posed by a major development in Vietnam. The results of that challenge demonstrated dramatically the extent to which the new trends had progressed, and at the same time demonstrated the attentuation of the Vietnam factor in world affairs. The event in question was the massive offensive launched at the end of March by the North Vietnamese in an all-out bid to resolve the impasse in negotiations by military force. That operation must have also been viewed by Hanoi as a forceful means of reasserting its interests with its big patrons, whose flirtation with the United States had been accompanied by the most acute symptoms of pain and misgiving on Hanoi's part. Now Hanoi was to witness gloomily how secondary its interests had become in the great-power calculus.

On 8 May 1972, President Nixon announced a move that the United States had avoided even at the high point of American military involvement: mining of Haiphong harbor and other North Vietnamese ports as part of an effort to interdict the sea and land supply routes. In his statement, Nixon addressed himself particularly to Moscow in an attempt to minimize the risks that heretofore had been regarded as prohibitive for such an operation. He pointed out that the Soviet Union and the United States had made "significant progress" in negotiations that now promised "major agreements" on nuclear arms limitation and other issues. Holding out the prospects of "a new relationship" serving not only the interests of the two countries but also the cause of world peace, he affirmed that the United States was "prepared to continue this relationship. The responsibility is yours if we fail to do so."[47] This was putting Moscow in a very delicate situation, as Henry Kissinger acknowledged at a press conference the next day when he said the administration's decision, taken "with enormous pain and great reluctance," presented "some short-term difficulties" to the Kremlin. Kissinger replied to a questioner that it was the administration's judgment that the mining decision did not pose an unacceptable risk, "especially compared to the risks of the situation where for the second time since the summit meeting was arranged Soviet arms fueled a military upheaval."[48]

The risk did, in fact, prove to be acceptable. Not only was the Moscow summit held as scheduled, but it produced the most euphoric expressions of cooperative intent yet, an intent expressed in particular in the first ever agreement by the two superpowers to limit strategic weapons. The two sides also joined in a declaration of principles over and above the joint communiqué on their talks. In addition to matching the Shanghai communiqué by identifying peaceful coexistence as the basis of their relations, the two sides went on to "attach major importance to preventing the development of situations capable of causing a dangerous exacerbation" of their relations.[49] The mining of Haiphong harbor, once regarded as one of the most provocative actions the United States could take in the face of Soviet support for the North Vietnamese, had hardly placed a strain on Soviet-U.S. relations, not to mention caused an exacerbation. As Nixon and Kissinger had indicated in their explanations of the situation arising from the Haiphong mining decision, Moscow and Washington had already reached the threshold of defining a new relationship, and this could not be lightly dismissed by the Soviets. A more overriding consideration for the Kremlin, however, was simply that the scheduled Moscow summit had been preceded by the one in China, and the Soviets could not afford to remain on the sidelines while new rules of the international political game were being evolved. In the 1960s, the Soviets could not afford to move too far ahead of Hanoi in trying to mediate an Indochina settlement, for Hanoi could play Peking off against Moscow. Now Peking perceived matters within an altered conceptual framework, and Moscow was compelled to keep pace even at Hanoi's expense. One consequence of the new realities was that Hanoi was constrained to separate the military and political aspects of its negotiating demands, severing a linkage to which it had tenaciously clung but which now no longer commanded essential support from its major allies. This paved the way for a Vietnam agreement that provided, in essence, for United States withdrawal along *with* the release of American POWs but *without* the dismantling of the Saigon government, thus realizing Hanoi's military but not its political demands.

The effects of the structural changes in the international system were most pronounced in the case of the Vietnam negotiations during the period between the 1972 summits and the signing of the agreement in January 1973. Where, in the past, pressure from Moscow could be offset and diverted by countervailing pressure from Peking, the systemic effect of the changes in adversary relations was that pressures from Moscow and Peking now reinforced rather than offset one another, and this resulted in intolerable pressures on Hanoi to adjust its negotiating position so as to offer a deal acceptable to Washington. Though Sino-Soviet rivalry had given Hanoi leverage and freedom of maneuver, the North Vietnamese were always concerned that this rivalry remain within certain bounds, lest their interests be lost in the larger conflict. Peking's provocative use of the border crisis had posed just such a danger, prompting Hanoi's earnest efforts to defuse that crisis. The trends that had clearly emerged by 1972 complicated Hanoi's task still further, as reflected in its efforts to come to terms with questions that, as the party organ *Nhan Dan* put it, had been "easy to answer in the past" but had now become "puzzling questions because of unhealthy tendencies."[50] One of the questions cited was the identity of the principal enemy, a question the Sino-Soviet border crisis and the two Nixon summits had posed to the Communist movement in the most fundamental terms. This was far more than a theoretical matter for the North Vietnamese, who now acknowledged with pained candor that they were being subjected to "terrible pressure" not only from the United States attacks and mining but also from "tendencies of compromise from outside."[51] Hanoi was, thus, expressing the most acute symptoms of the changes taking place in the international system.

During the months after the Moscow summit, the negotiating pace stepped up markedly, including a visit to Hanoi by Soviet President Podgornyy and one to Peking by Kissinger, in both cases natural sequels to the summit, and a succession of meetings in Paris between Kissinger and top North Vietnamese negotiator Le Duc Tho, whose shuttles between Hanoi and Paris included regular stopovers in Peking and Moscow. The gravity of the concern behind the shrill signals of alarm coming out of Hanoi was indicated by the use of the very serious charge of "opportunism" against those counseling compromise.[52] At this juncture, however, the "terrible pressure" from Hanoi's allies was reinforced by circumstances that must have sharply focused Hanoi's concentration on the timeliness of an agreement with the United States. First of all, of course, the military offensive launched that spring had been arrested well short of resolving the whole issue by force, thus leaving the question of a negotiated settlement in the forefront of consideration. Second, as in the case of the breakthrough four years earlier when the parties agreed to open negotiations, this was a U.S. presidential election year, which gave Hanoi the incentive to capitalize on American domestic pressure in order to nail down in negotiations whatever gains were now within reach but might recede after the election. Finally, the changed international environment created uncertainties, particularly concerning the disposition of Hanoi's major patrons, that argued for a move to take what could be gotten now and to remove the United States from the battleground, even if no substantial political cost could be exacted at the same time. This, then, was the context for the signal of a change in the Communists' negotiating position, conveyed in a statement by the Viet Cong on 11 September,[53] which coincided with the return to Paris once again of Le Duc Tho. The operative message in that statement was its assertion that a settlement must be founded on "the actual situation that there exist in South Vietnam two administrations, two armies, and other political forces." As spelled out in a draft agreement handed over by Le Duc Tho to Kissinger a month later, this proposal had as its crucial merit the necessary separation of the military and political issues. As late as September, an authoritative article in the theoretical journal *Hoc Tap* had explained Hanoi's tenacious insistence on the linkage of the military and political issues by observing that what the United States called military questions—such as a cease-fire, troop withdrawal, release of POWs—encompassed "a key political problem because it means recognition of the puppet administration" of Nguyen Van Thieu. *Hoc Tap* had also noted that the United States included among political questions one of a military character, namely "the question of the [North] Vietnamese armed forces in South Vietnam," and that it demanded a deferral of the political questions until some time after the settlement of what it defined as military questions.[54] The agreement reached was due to a compromise on these key elements: it was formally acknowledged that Vietnam was one country and therefore, as a corollary, that the North Vietnamese forces were not obliged to withdraw from the South; and the Thieu regime remained intact as a political entity, thus leaving the further evolution of the situation to the interplay of the Vietnamese forces.

The Vietnam agreement can be viewed as the fruition of a process that began in 1968 with the opening of negotiations in Paris. It can also, however, be viewed in the broader geopolitical context of the changes in the international environment that had as their impetus the initiatives undertaken by Peking in the wake of the invasion of Czechoslovakia, among which was a new openness to a Vietnam settlement that coincided with the reversal of the escalating U.S. involvement in that conflict. It was in the crucible of the Sino-Soviet border crisis that these initiatives acquired their transforming powers, leading to a conceptual change in Peking's view of the international system, and again, on this broader level, coinciding with a sharper perception by Washington of the possibilities of a new, evolving set of relationships in East Asia in place of frozen confrontation. In addition to the emergence of a new, conditionally cooperative relationship between the United States and the PRC, the age-old Sino-Japanese antagonism was replaced by a normalization of relations achieved within months of the Nixon China visit. It is fitting to recall that the 1950 Sino-Soviet treaty of alliance, as an updating of the wartime treaty between the Kremlin and the Kuomintang, was directed against Japan or "any state allied" with Japan, a formulation taking account of the postwar international environment and giving expression to Peking's decision to "lean to one side" in the bipolar conflict. That treaty had become a dead letter from a failure of mutual expectations of the alliance partners, and now Peking had developed relations with former adversaries in order to check what it perceived to be the hegemonial drive of its erstwhile ally.

THIRTEEN

Concluding Propositions

This study of political change has focused on the Sino-Soviet border crisis as symptomatic of structural shifts in the alliance and adversary relations shaping the international system. The little more than a year between the Soviet invasion of Czechoslovakia in August 1968 and the opening of Sino-Soviet border negotiations in October 1969 witnessed the eruption of tensions that severely tested the stability of the international system, but it was also a period of the genesis of significant political initiatives that began to take concrete form after the Sino-Soviet confrontation had been brought into an uneasy equilibrium and the stage was prepared for a dynamic Chinese thrust into diplomatic activity. Thus, there were systemic changes consisting essentially of a transformation of adversary relations in which the erstwhile partners forming the axis of "the socialist camp" had become primary enemies, while a conditionally cooperative relationship replaced the long-frozen adversary relationship between the United States and China. The rapidity with which an antagonistic confrontation of such long standing, and of such seeming intractability, could be transformed was testimony to the fundamental nature of the underlying forces of change. Likewise, the swift turnabout in Sino-Japanese relations reflected the new fluidity imparted by this basic change. Above all, these developments reflected the primacy of the Sino-Soviet conflict as the locomotive of structural change. Nonetheless, for all that these developments, and their antecedents in successive phases of Sino-Soviet conflict since the Soviet 20th Party Congress in 1956, testified to the gravely troubled nature of the Sino-Soviet relationship, a contextual analysis of the events constituting these developments has shown how inherently political a process it has been. In this sense, there is only limited analytical utility in probing the "deep" historical and geopolitical roots of those troubles. They can be found embedded in the shape and workings of the postwar international system, and it is in that context that we should look for the conclusions to be drawn from this study.

The border crisis was essentially political. There is a virtually tautological sense in which the Sino-Soviet border conflict is what it is because of the historical past; but history and geography do not in themselves account for the politics of the border crisis, or for the uses the parties involved have made of the border dispute as an instrument of the Sino-Soviet rivalry that began to take definite form at the beginning of the 1960s. When the Sino-Soviet alliance became troubled, for more inherently political reasons than unresolved territorial issues, the border became troubled. One of the most sensitive symptoms of mistrust and suspicion is a territorial dispute, touching, as it does, on such fundamental concerns as territorial integrity and national identity, and this was reflected in the emergence of a border conflict between Moscow and Peking as their rift opened and mutual mistrust developed during the decade. As discussed in Chapter 4, the Chinese elected to make an issue of border tensions in the context of the invasion of Czechoslovakia, and did so for political purposes in their challenge to the bipolar international system that they bitterly perceived as having been demonstrated in the case of Czechoslovakia.

What was striking about the emergence of a border confrontation in the wake of Czechoslovakia was the relative quiescence of the border issue in the immediately preceding years, which had witnessed severe disturbances in Sino-Soviet relations in practically every other sphere. It would be expected that, during periods of disintegration of central authority, and of a witch-hunt atmosphere making counsels of restraint hazardous, as during the Chinese Cultural Revolution, a festering territorial dispute would be enflamed and be likely to flare out of control. In fact, however, it was not until after the lines of authority, of political command and control, had been reconstituted in China that the border crisis emerged. As the chronology of events discussed in Chapter 4 shows, the moves taken by the Chinese appeared to be well calculated and controlled by central authority. Moreover, the early moves on the border issue in the aftermath of Czechoslovakia were taken during a period of notable political initiatives being undertaken by Peking. The significant turnabout in Peking's posture on the Vietnam negotiations, together with the overture to the incoming Nixon Administration to open a political dialogue on basic issues, coincided with the appearance of a supple, differentiated East Europe policy that testified to a confident new political decision-making environment in Peking. It was in such an environment that the border issue could be pursued and manipulated as a political instrument.

Care must be taken in relating initiatives in the international political arena to domestic politics and its contending factions. The Sino-Soviet relationship has been an important element in Chinese leadership politics (much less so in Soviet internal politics), but this cannot be reduced to simple categories of pro- and anti-Soviet or degrees of antipathy to Moscow. Rather, a more complex and concrete analysis is required which takes into account the differential effects on the positions and priorities of various leaders and groups of various policy options and initiatives. The initiatives undertaken by Peking in the wake of Czechoslovakia, which included the airing of border tensions, represented departures from the dual confrontationist policy that had dominated Peking's approach since 1965, the time of the debate over the implications of Vietnam and the post-Khrushchev leadership's offer of united action. Moreover, these moves coincided with an important juncture in domestic affairs, the completion of the rebuilding of provincial revolutionary committees and the convening of a Central Committee Plenum that put a new party congress on the agenda. These developments, marking, in a sense, the winding up of the Cultural Revolution, raised the question of the future position and fortunes of those elements, most notably Lin Piao, that had risen to power with the Cultural Revolution. It is in this context that the fanfare surrounding the anniversary in September 1968 of a hitherto unpublished Mao inscription acquires significance as a kind of warning signal that the Cultural Revolution was ending and its beneficiaries' future might be open to some doubt.

Viewed in this context, the border crisis as such did not represent one faction's anti-Soviet orientation or a move by one faction to capitalize on the border issue at the expense of others. That crisis was symptomatic of the underlying changes resulting from the invasion of Czechoslovakia, and so was part and parcel of the new turn taken by Chinese policy during this period; however, it was the broader uses of the border conflict that determined its implications for Chinese domestic politics. That is to say, the anti-Soviet animus of much of Chinese policy was not a differentiating factor in leadership politics, but the direction given to that animus within a wider policy context was such a factor. For Chou En-lai and his associates the border issue served as an instrument in turning Chinese policy from the dual confrontationist track of Lin Piao to a more differentiated approach in which the United States was perceived as a secondary adversary and, to some extent, as a partner in "parallel" policies vis-à-vis Soviet hegemonial impulses. Like the move, after the Chinese Party Congress in April 1969, to rebuild the party apparatus, this Chouist policy approach posed a challenge to powerful interests, notably Lin and his associates, whose power and positions were, perforce, threatened by change. The containment of the border crisis and the move to a negotiating track after the long hot summer of 1969 marked (contrary to Moscow's hopes) not a reduction in the anti-Soviet animus but a more differentiated use of the border conflict in the interest of a sharply revised strategy. As reflected in the revealingly polemical *Red Flag* article appearing immediately prior to the invitation to Kosygin to confer with Chou in Peking, the change in strategy had important implications for Chinese leadership politics as well as for world politics.

In a very broad sense, the change in Moscow's China policy after the ouster of Khrushchev was comparable to the later change of strategy by Peking. Like Lin Piao, Khrushchev had represented an aggressively uncompromising posture, while Khrushchev's successors, similar to the Chouist approach later, seemed to hold open the prospect of a political-diplomatic approach to reduce Sino-Soviet tensions. In each case, however, the change was toward a more effectively flexible position in the international arena, reversing a process of declining influence. Thus, the post-Khrushchev leadership shelved the divisive project for an international party conference ostracizing the Chinese and advanced the line of united action that proved effective in restoring the badly eroded Soviet influence in the Asian Communist movement. Moscow also took care to develop a strong military presence along China's borders, but its main gains were political as Peking retreated into the angry isolationism of the Cultural Revolution. This left the field largely to Moscow in the late 1960s, until the Chinese policy turn at the end of the decade thrust Peking back into the diplomatic picture and registered impressive political gains (for example, the opening to the United States, seating in the UN, widespread diplomatic recognition) in the first years of the new decade. Amid these political changes, however, the essentials of the territorial dispute as such remained unresolved.

The border crisis was primarily a Chinese initiative. It was, after all, Peking's decision to air border tensions and to press an issue that had remained largely under cover. Peking related this decision to Czechoslovakia, and it used the issue in the service of its strikingly new East Europe policy that emerged from the invasion of Czechoslovakia. In particular, as discussed in Chapter 4, the border issue was interwoven with the campaign of Sino-Albanian mutual security, the centerpiece of the "historic new stage" of struggle proclaimed by Peking in September 1968. This new stage emerged from strategic considerations, as distinguished from the ideological interests binding Peking and Tirana in the previous phases of their close relationship; in a broader context, this was reflected in Peking's support for Romania against the Soviet menace and in the ensuing close Sino-Romanian relationship, but perhaps even more tellingly it was reflected in the "negative signal" of Peking's cessation of its ideological and political campaign against Yugoslavia. The Chinese followed up the polemical use of border tensions by aggressive patrolling of disputed areas, thus setting the collision course that issued in the Chenpao/Damanskiy Island clash along the icebound Ussuri. The Chinese seemed intent upon demonstrating Moscow's "social-imperialist" impulses, and, in a sense, they were provoking those impulses, the aggressive patrolling of disputed areas being a most effective way to do so.

For their part, the Soviets were extremely loath to acknowledge the border issue as a juridical question and were hopeful that a combination of political restraint and military firmness would restore the rules of the game that had previously prevailed along the border. This hope was reflected in Moscow's burying its protest note on the 2 March 1969 clash in *Pravda's* inside pages and its conspicuous failure to match the massive Chinese political and propaganda campaign on the matter for several days thereafter. When the Soviets finally did mount a political counteroffensive, *Pravda* discerned "far-reaching aims" being served by Peking's aggressive moves, but the territorial question was given only passing mention. The second clash, on 15 March, reflected the Soviet effort to demonstrate military firmness as a way to deter the Chinese, and it was not until two weeks later still that Moscow tried to reopen the 1964 border talks as a means of defusing the crisis the Chinese had seemed so intent on provoking. The series of Soviet steps in this direction, with their attendant concessions on Moscow's part, testified to the extent to which Moscow was the reactive party trying to stabilize a situation fraught with unforeseeable consequences:

—Kosygin tried to telephone Peking on 21 March to discuss the crisis, but he was rebuffed by the Chinese.

—Moscow issued a Government Statement on 29 March proposing resumption of the 1964 "consultations" on the border question.

—While the Chinese Party Congress was in session in April, Moscow issued a note free of polemics appealing to Peking to resume the border talks. Moscow followed up with a week-long moratorium on anti-Chinese polemics.

—Moscow, on 26 April, proposed holding what until 1968 had been an annual session of the joint commission on border river navigation. The Soviets hoped to agree on a modus operandi with the Chinese that would free the approaching shipping season from the tensions and clashes that had marked the icebound season along the rivers.

—The Soviets continued to fire on Chenpao but did not try to occupy the island.[1]

—After the Chinese had belatedly, two months later, replied to the Soviet offer to reopen talks on the border, Moscow, on 13 June, issued a statement welcoming the indication in the Chinese reply that necessary adjustments along particular stretches of the border could be made on the basis of existing treaties. Moscow named a chief delegate to the proposed talks and asked that they begin in two or three months.

—In an authoritative assessment of the anti-Chinese International Party Conference held in Moscow in June, Brezhnev assumed a posture of "calm and restraint" while giving the assurance that the Soviets would not be provoked into rash actions.

—Moscow swallowed its pride and had Kosygin double back to Peking after reaching Soviet Central Asia on his return from Ho Chi Minh's funeral in Hanoi. Moscow lent a positive cast to the Kosygin-Chou meeting and immediately entered into a polemical stand-down. The Soviet message on the PRC National Day called for a settlement of conflicts through "consultations and negotiations."

—Instead of reopening "consultations" in Moscow with the border guards chief as the Soviet delegate—which had been the terms proposed by Moscow to show continuity with the 1964 consultations and to avoid any impression of renegotiating the border treaties—the two sides agreed to hold border negotiations in Peking at the level of deputy foreign ministers.

Due appreciation of Chinese moves to press the border issue and of Soviet steps to defuse the crisis is an important corrective to facile dramatic accounts that overemphasize Soviet military pressure and threats. It was Moscow that repeatedly made the concessions, and the Chinese remained in possession of the disputed island that had sparked the fighting. Another source of oversimplified accounts of the border crisis has been a tendency to see the Chinese moves in the wake of Czechoslovakia as being responsive to a Soviet threat to China. Such a view has Peking reacting out of apprehension over the implications of the Brezhnev Doctrine generalizing the Soviet right of intervention in any Communist country, presumably including China. But that doctrine was enunciated to cover a particular intervention, and there was nothing about developments in China at that particular time which would make it more likely that the Soviets would move forcibly to change the Peking regime. Moreover, if the Chinese were fearful of any such Soviet move, they would hardly choose to be provocative at that particular time by raising such a sensitive issue as the border question and to challenge the rules of the game that had contained tensions along the border. The military dimension of the border crisis must be kept in its place, subordinate to the political and strategic considerations that underlay Peking's policy initiatives following the invasion of Czechoslovakia.

The postwar international system provided the matrix for the Sino-Soviet conflict, including the border crisis. This proposition amplifies the one asserting the essentially political nature of the border crisis. It is analytically rather vacuous to view the Sino-Soviet conflict as being the ineluctable outcome of deep historical forces in the Russian and Chinese pasts, from which the border crisis arose as a well-nigh inescapable collision of two imperial territorial imperatives in contest for centuries. That conflict has been rooted in the structural character of the postwar

Concluding Propositions 277

international environment and its effects on the politics and perceptions of the Soviet and Chinese Communist regimes. The postwar system was the direct result of World War II, and (tautologically) that war was the product of the historical forces leading to it; however, World War II was not the product of forces intrinsic to the historical background of relations specifically between Russia and China. One might look at the matter this way: in the complex and varied causes of World War II, the historical relations between Russia and China have only a tangential place.

As discussed in the Introduction, the central structural characteristic of the postwar international system was its bipolarity, which set the framework for Sino-Soviet relations within the pattern of alliance and adversary relationships shaped by that bipolarity. After the watershed developments of the late 1950s, when fundamental divergences in perceptions and expectations concerning the uses of the Sino-Soviet alliance emerged in sharp relief, Peking began to take significant moves away from integration within the bipolar system and toward greater Chinese independence, a direction that prompted Soviet apprehension over Chinese acquisition of nuclear weapons and induced Moscow to pursue arms control agreements with the United States over bitter Chinese protests. The exigencies of the Vietnam War, interacting with the fierce insistence on independence by the ascendant Maoist-nationalist wing of the Chinese leadership, drove this secular trend into a deeply isolationist byway. During this isolationist phase, the Chinese in effect withdrew from the international system and had recourse to extra-systemic means to influence foreign affairs; this was the meaning of the doctrine of people's war enunciated by Lin Piao in 1965 as the United States intervention in Vietnam was intensifying and the Chinese rejected socialist camp unity. Peking's assessment of the implications of Czechoslovakia, seen through the Vietnam prism, served as the starting point of still another set of initiatives (Chapters 4 and 5) which, as they began to take shape in the 1970s, had a transforming effect on the international system. The border crisis was an acute symptom of the destabilizing effects of this structural change, as the Hungarian revolt in 1956 had manifested the structural tremors resulting from Khrushchev's attempts to adjust international Communist relations to accommodate his post-Stalinist initiatives. But from the limited exercise of carefully controlled force that created the border crisis there emerged a new dynamic equilibrium that represented a transformed international system, one in which Peking found instrumentalities for breaking through the bipolarity with which the Chinese had been trying to come to terms for so many years. China now proceeded toward reintegrating itself into the international system. The rapid development of diplomatic and political relations across a broad global front, the PRC's seating in the United Nations, the engagement of the United States and Japan in political-diplomatic and economic relations to offset Soviet influence in East Asia, the insistent promotion of countervailing forces against the Soviet Union (especially in Europe)—all of this testified to the extensive reintegration of China in the international system following the border crisis.

The Taiwan question played a subordinate role in the emergence of a new Sino-American relationship. Since the establishment of the PRC, the Taiwan issue has been a central concern of Peking's. Several important results would turn on a successful outcome of Peking's repeated efforts, using varying methods, to acquire control over Taiwan: consummation of the revolution that was halted in 1950 when the outbreak of the Korean War led the United States to interpose itself between the mainland and Taiwan; removal of a threat to the PRC's security posed by a nearby highly armed enemy; and elimination of a living challenge to Peking's political authority. The U.S. interdiction of the Taiwan Straits at the outset of the Korean War froze Sino-American relations into an adversary relationship of fundamental proportions that wellnigh defied real change for two decades. The Taiwan Straits having become one of the lines of great-power conflict, the Taiwan issue also became a key factor in alliance politics, with a particularly corrosive effect on the Sino-Soviet relationship. All of this notwithstanding, the Taiwan question had a subordinate part in the basic changes that had their source in the late 1960s and began to reshape the international political environment in the next decade. This was so because it was Soviet, not American, behavior that was Peking's primary concern underlying the initiatives inducing those changes. Progress on the Taiwan issue was at once a result and an enabling condition of the more basic changes, but the fact remains that the latter were primary while movement on the Taiwan question was derivative.

The Taiwan question had figured in the overture to the United States in November 1968 that was one of the series of initiatives undertaken by Peking in the wake of Czechoslovakia. At that time, with its call for U.S. military withdrawal from Taiwan and for an agreement on Sino-American peaceful coexistence, Peking had insisted that the two sides tackle fundamental issues and not engage in "haggling over side issues." With this insistence and its positing of peaceful coexistence as a goal of negotiations, Peking signaled that the new directions in its policy were pointing to fundamental changes in the international political environment, changes that for Peking turned essentially on identifying and enlisting countervailing forces against the Soviet hegemonial thrust that the Chinese perceived as the meaning of Czechoslovakia. The single most important source of such countervailing forces being the United States, Peking was preparing the ground for a transformation of the adversary relationship that had been an enduring legacy of the Korean War but that now, for Peking, was ripe for rethinking in view of the primacy being assigned to the Soviet challenge. (Likewise for the United States, envisioning a post-Vietnam situation, the time was ripe for a change in the Sino-American relationship.) An accommodation on the Taiwan issue was a necessary element in this change; the purpose of the change was not to solve the Taiwan question but, rather, an accommodation on the Taiwan question was for the purpose of more fundamental structural change in the international environment. This ordering of priorities and purposes was confirmed by the results of the February 1972 Sino-American summit. The two sides agreed to base their relations on peaceful coexistence, they jointly declared their opposition to Soviet hegemony, but they accepted a compromise formula for dealing with the crucial issue in their bilateral relations, Taiwan. As has been seen, Peking reiterated its demand that the United States withdraw from Taiwan, while the U.S. conditioned its commitment to withdrawal on a peaceful settlement of the question by the Chinese on both sides of the Straits. The issue remained an open one, but the accommodation that had been reached was a necessary enabling condition for the transformed Sino-American relationship.

The signaling process is crucial to understanding political change. The key analytical concept in this study of political change has been that of political signaling. If nothing else has been achieved, it is hoped that the simplistic and analytically pernicious word/deed dichotomy—along with such variants as thought/action and intention/capability—has been eroded. Signaling is a form of communication having *operative* meaning: to say certain kinds of things in certain ways in certain contexts is to *do* certain things, such as to create expectations, convey threats, offer promises, undertake commitments, prescribe attitudes, identify friend and foe; in short, to initiate or induce change by influencing others' expectations, intentions, and perceptions.

One source of misunderstanding is the failure to distinguish the logical from the syntactical form of signals. Syntactically, these communications may have the form of descriptive or representational statements: they appear to depict or represent a state of affairs or facts. The logic of meaning, however, relates to what these statements *do*, what function they serve, the role they play: their operative meaning is to signal something, which is intrinsically a political process in seeking to influence expectations, intentions, and perceptions. The logical condition of "esoteric communication" lies in an extreme dissociation of operative meaning and syntactical form. Likewise, a seemingly innocuous statement, or indeed a mere phrase, may have portentous meaning in its political as distinguished from its syntactical context. (The latter context is the full text; content analysis deals with texts.) Its operative meaning as a signal has an internal relation to the political context, and it is the purpose of contextual analysis to apprehend that meaning as a significant factor in the political process. It is also true, of course, that failure to apprehend signals, or their misapprehension, can be a decisive factor in the political process, and this too is an important subject of contextual analysis.

The counsel to "watch what they do, not what they say" has often been a source of misunderstanding, or of "under-understanding." After the record of Chinese restraint and very low-risk involvement in the Vietnam War had been established in the late 1960s, one often heard this cited as a case in point, confirming the need to watch what they do rather than attending to their bellicose rhetoric. But, already in 1965, Peking had signaled its intention to pursue the isolationist, low-risk policy maintained in the following years: this was the operative meaning of the statement of strategy conveyed in Lin Piao's notorious people's war tract. A different sort of example was the reaction in the West to the crisis arising from the fighting along the Sino-Soviet border in 1969. Those who would put little store in the signals of the preceding autumn—if indeed they even noted them—found the dangers in the summer of 1969 of large-scale hostilities to be quite high, the estimate being that the Soviets were likely to escalate the level of hostilities by some sort of preventive strike. Then, when the crisis was instead defused and negotiations undertaken, the simplistic explanation was that Soviet intimidation had driven the Chinese to the negotiating table. Such an analytical approach, or abdication of an analytical approach, failed to comprehend the signaling process and the overall political context in which those events took place.

Ultimately, the political context in which signals and perceptions are meaningful is a conceptual context. Misperception or failure to apprehend a signal as likely testifies to a conceptual fault as to inattentiveness or obtuseness or lack of information. Accordingly, this study has sought to put the Sino-Soviet relationship in conceptual context by reference to the postwar bipolar international system, and to the interplay of Soviet and Chinese perceptions and signaling within that context. For the Chinese, the

Soviet invasion of Czechoslovakia posed a challenge that at bottom was a conceptual challenge, one of a series, but a critical addition to the series, of challenges to a system (of which "the socialist camp" was a key sub-system) which had produced expectations that the bipolar structure could not accommodate. Hence, Peking initiated a signaling process that, most importantly, soon coincided with a move in the same direction by the United States. The result was a conceptual change in outlook, in which each side not only perceived the other in a fundamentally new way but also helped create one another's perceptions, and with this process came the system transformation that has been the subject of this study.

Though this has been a retrospective study, it is the real task of contextual analysis to monitor political communications for semiotic alerts, or early warning signals, of political change. Two qualifications are required here. First, these communications may not have the conventional forms of political discourse: esoteric political communication, for example, has taken such diverse forms as historiography or literary criticism or abstract philosophy. Second, the signaling process should not be regarded as synonymous or coterminous with political communication as such. Though every political statement may to some degree have a signaling aspect, some forms or contexts of communication are essentially or particularly used in a signaling process. On one end of the spectrum would be straightforward announcements and directives, or propaganda and publicity, which convey information or seek to rally support. In these types of communication the operative meaning is what is being said in a straightforward sense. At the other end of the spectrum, as in esoteric communication, there is that dissociation of operative meaning and the straightforward content of the communication that is the logical condition for much of the richness and subtlety of the signaling process. This is not to say, however, that the signaling process relies mainly or paradigmatically on allegorical communications: manipulations of syntactical form, as in subtle reformulations, or omissions, or new emphases, can also serve the process with telling effectiveness.

Moreover, the simple reappearance of an old formulation (or conversely, the disappearance of a standard one), interpreted in context, may be seen as a signal of something new or as initiating a signaling process. Sometimes the most straightforward factual report, or a juxtaposition of such reports, can be loaded with operative meaning as a signal when interpreted contextually. The forms signaling can assume are endless, which makes a taxonomy incomplete at best, for signals acquire their meaning contextually, having an internal relation to the changing political context from which they are inseparable.

It is particularly when continuities of policies or patterns of relationships are in question or under challenge or undergoing significant modification that a signaling context emerges. This is a dynamic process in which expectations and perceptions are created that, in turn, are constitutive of others' expectations and perceptions through the signaling feedback. When established patterns or states of relative equilibrium become subject to serious challenge or disturbance, a crisis can emerge as the tensions of structural change become severe and concentrated in foreshortened perspectives. This is what happened in the Sino-Soviet conflict within the context of the signaling process originating in the wake of the Soviet invasion of Czechoslovakia. It is what produced the politics of the border crisis. It is why that political crisis was symptomatic of a change in the international political environment.

Notes

Bibliography

Index

Abbreviations

ABM	anti-ballistic missile
CCP	Chinese Communist Party
CPSU	Communist Party of the Soviet Union
CPUSA	Communist Party of the United States
DRV	Democratic Republic of Vietnam
DPRK	Democratic People's Republic of Korea
FBIS	Foreign Broadcast Information Service. References are to the various geographical area volumes of the *FBIS Daily Report* and to the two analytical reports, *Survey of Communist Propaganda* and *Trends in Communist Propaganda*.
GDP	General Political Department (PRC)
ICBM	inter-continental ballistic missile
JPRS	Joint Publications Research Service
KCNA	Korean Central News Agency (Pyongyang)
MPR	Mongolian People's Republic
NCNA	New China News Agency (Peking)
NPC	National People's Congress (PRC)
PLA	People's Liberation Army (PRC)
PR	*Peking Review*
PRC	People's Republic of China
ROC	Republic of China
SALT	Strategic Arms Limitation Talks

Notes

CHAPTER ONE

1. This section draws on Richard Wich, "Chinese Allies and Adversaries," in William W. Whitson, ed., *The Military and Political Power in China in the 1970s* (New York, 1972), pp. 291–312.
2. See Soviet Government Statement, 21 September 1963, in William E. Griffith, *The Sino-Soviet Rift* (Cambridge, Mass., 1964), pp. 426–461.
3. J. L. Austin, *How To Do Things With Words* (New York, 1970), pp. 1–11.

CHAPTER TWO

1. Edgar Snow, *Red Star Over China* (New York, 1944), p. 96n.
2. Dennis J. Doolin, *Territorial Claims in the Sino-Soviet Conflict* (Stanford, 1965), p. 43.
3. *Pravda* editorial article, 2 September 1964, "In Connection With Mao Tse-tung's Talk With a Group of Japanese Socialists," in Doolin, *Territorial Claims*, p. 52.
4. For the Chinese version, see *People's Daily* and *Red Flag* editorial departments, "The Origin and Development of the Differences Between the Leadership of the CPSU and Ourselves—Comment on the Open Letter of the Central Committee of the CPSU (1)," in Griffith, *Sino-Soviet Rift*, pp. 388–420. For the Soviet version, see Soviet Government Statement, 21 September 1963, in ibid., pp. 426–461.
5. N. Khrushchev address to Supreme Soviet, Moscow, 12 December 1962, in Doolin, *Territorial Claims*, pp. 27–28.
6. CPUSA statement, 9 January 1963, in ibid., pp. 28–29.
7. *People's Daily* editorial, "A Comment on the Statement of the Communist Party of the U.S.A.," in ibid., pp. 29–31.

8. "Open Letter from the CPSU Central Committee to Party Organizations and All Communists of the Soviet Union," 14 July 1963, in Griffith, *Sino-Soviet Rift*, pp. 289-325.
9. See the Chinese article, "Origin and Development of Differences," cited in note 4 above.
10. See the Soviet statement cited in note 4 above.
11. See the articles in *Kazakhstanskaya Pravda* by Usman Mametov, 22 September 1963, O. Matskevich, 24 September 1963, and Zunun Taipov, 29 September 1963, cited in Griffith, *Sino-Soviet Rift*, p. 15 n. 7.
12. This propaganda reappeared in early 1967, a time of accelerated deterioration of Sino-Soviet relations, after having been avoided in the first two years after Khrushchev's ouster. For the Sinkiang People's Congress session, see Doolin, *Territorial Claims*, pp. 40-41.
13. Soviet Government Statement, 21 September 1963, cited in note 4.
14. CPSU Central Committee letter to CCP Central Committee, 29 November 1963, in William E. Griffith, *Sino-Soviet Relations, 1964-1965* (Cambridge, Mass., 1967), pp. 147-152.
15. Khrushchev letter to heads of state or government, 31 December 1963, in Doolin, *Territorial Claims*, pp. 33-36.
16. CCP letter to CPSU, 29 February 1964, in Griffith, *Sino-Soviet Relations*, pp. 181-190.
17. *Peking Review* 7.18:6-12 (1 May 1964).
18. This distinction between inherited boundary questions and "imperialist" occupation provides a background for Peking's later charges, after the invasion of Czechoslovakia, of Soviet "social-imperialist" behavior and violations of the Chinese border.
19. Mao statement in Doolin, *Territorial Claims*, pp. 42-44.
20. *Pravda* editorial article in ibid., pp. 47-57.
21. Khrushchev interview with Japanese delegation, 19 September 1964, in ibid., pp. 68-72.
22. Gromyko letter to U.N. Secretary General U Thant, 21 September 1964, in ibid., pp. 72-74.
23. See the Chinese letter cited in note 16 above.
24. As Maxwell makes clear in his study of the Sino-Indian border conflict, Peking's treatment of the border question in the eastern sector was essentially determined by a desire to bargain its acquiescence in India's insistence on the McMahon Line in exchange for Indian acceptance of a boundary line in the west that would confirm China's title to the territory through which the Sinkiang-Tibet road runs. Neville Maxwell, *India's China War* (Garden City, 1972), pp. 162-164.
25. Ibid., p. 98.

Chapter Three

1. In fact, the conference project was a revival of Khrushchev's anti-China conference plan that had been shelved in March 1965, a few months after Khrushchev's downfall.
2. That debate has been the subject of extensive exegesis. For example, see Wich, pp. 297–303.
3. In an even more telling sign of intransigence, the Chinese not only refused to acknowledge publicly the opening of talks but censored Hanoi's own references to them when reporting North Vietnamese commentary.
4. FBIS *Survey of Communist Propaganda*, 29 August 1968, pp. 1–2, 12–14. Though with evident reluctance and painful second thoughts, the Albanians put the interests of mutual security against the Soviets above their bitter grievances against the Yugoslavs—a measure of the transforming effects of Czechoslovakia. Tirana for a while ceased its steady drumbeat of polemics against the Yugoslavs after the invasion. Reflecting Albanian leader Hoxha's venomous quarrel with Tito, however, polemics were again unleashed on the occasion of Hoxha's birthday on 16 October, and sensitivity to "Soviet and Western charges" that Tirana and Peking had taken positions on Czechoslovakia similar to those of the "Titoists and imperialists" elicited a stinging indictment of "Titoite treachery" and a defense of Albanian ideological purity by the party organ *Zeri I Popullit* on 5 November. It was revealing, nonetheless, that Tirana did not exploit U.S. Under Secretary of State Katzenbach's mid-October visit to Yugoslavia, something which, before Czechoslovakia, would have provoked diatribes against the Tito regime as the Trojan Horse of the Communist movement.
5. FBIS *Survey*, ibid., pp. 18–20.
6. FBIS *Daily Report: Latin America*, 26 August 1968, 01–24. FBIS *Survey*, ibid., pp. 21–27.
7. KCNA, 22 August 1968, in FBIS *Asia & Pacific*, 22 August 1968, D1–4.
8. *Nodong Sinmun* editorial, 23 August 1968, in FBIS *Asia & Pacific*, 23 August 1968, D1–9.
9. The Hanoi radio broadcast the preface and the TASS announcement on 21 August. The Hanoi press carried these on the next day, and on the 23rd *Pravda* published Hanoi's preface. See FBIS *Survey*, 29 August 1968, pp. 31–32.
10. *Pravda* editorial article, 22 August 1968, "The Defense of Socialism Is a Supreme International Duty," in FBIS *Soviet Union*, 23 August 1968, A14–36.
11. Pham Van Dong speech in FBIS *Asia & Pacific*, 3 September 1968, K1–13.

12. NCNA, 10 August 1968, in FBIS *China*, 12 August 1968, A1–6.
13. *Zeri I Popullit* editorial article, "The Defeat of the Soviet Revisionists in Bratislava," in FBIS *Eastern Europe*, 12 August 1968, B1–4.
14. *People's Daily* "Commentator," "Total Bankruptcy of Soviet Modern Revisionism," in FBIS *China*, 23 August 1968, A5–8. NCNA report in ibid., A1–5.
15. FBIS *China*, 26 August 1968, A1–3. Chou's speech and the 23 August "Commentator" article were published in a special supplement to the issue of *Peking Review* released on that date. This reflected Peking's concern to get its message out promptly and widely. *PR* 11.34: Supplement (23 August 1968).
16. *People's Daily* "Commentator," 30 August 1968, in FBIS *China*, 30 August 1968, A1–3.
17. *PR* 11.36:12 (6 September 1968).
18. *People's Daily* and *Red Flag* editorial departments, "Refutation of the New Leaders of the CPSU on 'United Action'," in *PR* 8.46:10–21 (12 November 1965).
19. Moscow has varying uses for the two terms. Briefly, "the socialist community" refers to a network of political, military, economic, ideological, and other relations binding Moscow and its allies, whereas "the socialist camp" connotes mainly mutual defense relations in a bipolar world.
20. Chou En-lai's speech on Democratic Republic of Vietnam National Day, 2 September 1968, in *PR* 11.36:6–7 (6 September 1968).
21. *PR* 11.37:31–32 (13 September 1968).
22. *PR* 11.39:28–29 (27 September 1968).

Chapter Four

1. *PR* 11.38:41 (20 September 1968).
2. The NCNA international service introduced its account of the article with this warning. FBIS *China*, 23 September 1968, A1–2.
3. The Chinese message in *PR* 11.38:3–4 (20 September 1968); for a Chinese account of the Albanian law, see ibid., p. 7, and Shehu's address, pp. 8–14.
4. *People's Daily* editorial in *PR* 11.39:9–10 (27 September 1968).
5. *People's Daily/Liberation Army Daily* joint editorial, "A Great Fighting Friendship," in FBIS *China*, 30 September 1968, A13–16.
6. Ibid., 1 October 1968, A1–3.
7. Huang and Balluku speeches in *PR* 11.41:7–11 (11 October 1968).
8. *PR* 11.40:14–15 (4 October 1968).
9. FBIS *China*, 8 October 1968, A3–4.

10. Peking had begun broadcasting in Serbo-Croatian in June 1961 and in Russian in March 1962, thus providing channels for the polemical war then being waged against "modern revisionism."
11. Even that 1967 trip was very much a product of the Red Guard era rather than a diplomatic exercise, having been a visit by a Red Guard delegation under Cultural Revolution ideologue Yao Wen-yuan to attend an Albanian youth congress. Huang's delegation represented a move out of the isolation of the Cultural Revolution era and into an era in which strategic rather than ideological considerations prevailed.
12. *PR* 11.50:7-9 (13 December 1968).
13. FBIS *Soviet Union,* 31 October 1968, A12-13.
14. Ibid., 5 November 1968, A10-11.
15. The Russian edition was dated 14 February 1969, the anniversary of the Sino-Soviet treaty of alliance, which for the third consecutive year went unmarked by either side.
16. V. Shelepin, "Albania in Peking's Plans," *New Times* (English edition) 7:19-20 (19 February 1969).

CHAPTER FIVE

1. *People's Daily/Liberation Army Daily* joint editorial, "Long Live the All-Round Victory in the Great Proletarian Cultural Revolution," in *PR* 11.37:3-5 (13 September 1968).
2. NCNA, Urumchi, 6 September 1968, in FBIS *China,* 9 September 1968, H1-4.
3. NCNA, Lhasa, 6 September 1968, in ibid., E1-4.
4. *People's Daily/Red Flag/Liberation Army Daily* joint editorial, 1 October 1968, in *PR* 11.40:18-20 (4 October 1968).
5. Plenum materials in *PR* 11.44 Supplement (1 November 1968).
6. *PR* 11.38:2 (20 September 1968).
7. *People's Daily* editorial, 18 September 1968, "Compass for the Victory of the Revolutionary People of All Countries," in ibid., pp. 5-6.
8. Philip L. Bridgham, "The International Impact of Maoist Ideology," in Chalmers Johnson, ed., *Ideology and Politics in Contemporary China* (Seattle, 1973), p. 342.
9. *People's Daily/Red Flag/Liberation Army Daily* joint editorial, 1 January 1969, "Place Mao Tse-tung's Thought in Command of Everything," in *PR* 12.1:7-10 (3 January 1969).
10. NCNA, 19 October 1968, "Vietnam-U.S. 'Paris Talks' Enter 'Delicate Stage'," in *PR* 11.43:12 (25 October 1968).
11. NCNA, 8 November 1968, "Swapping Horses in Turbulent Waters," in *PR* 11.46:28-29 (15 November 1968).

12. NCNA, 21 November 1968, in *PR* 11.48:26-28 (29 November 1968).
13. Ibid., pp. 30-31.
14. *PR* 8.8:12 (19 February 1965).
15. Texts of the Chinese ambassador's statement and press conference in Warsaw, 7 September 1966, in *PR* 9.38:7-10 (16 September 1966).
16. Mao's 1949 report in *PR* 11.48:3-9 (29 November 1968); *People's Daily/Red Flag/Liberation Army Daily* joint editorial, 25 November 1968, "Conscientiously Study the History of the Struggle Between the Two Lines," in ibid., pp. 10-13.
17. *PR* 12.5:7-10 (31 January 1969). See FBIS *Survey*, 30 January 1969, pp. 1-6.
18. NCNA, 28 January 1969, in FBIS *China*, 29 January 1969, A1-2.
19. NCNA, 1 February 1969, in ibid., 3 February 1969, A3-4.
20. Ibid., 23 January 1969, A1-3.
21. *PR* 12.5:5-7 (31 January 1969).
22. *PR* 12.7:20 (14 February 1969).
23. Hung Tsai-ping, "Clumsy Performance," *People's Daily*, 18 February 1969, in FBIS *China*, 19 February 1969, A1-2.
24. Ibid., 18 February 1969, A8.
25. See *People's Daily* article by Hung Hsueh-ping reported by NCNA, 5 December 1968, in ibid., 5 December 1968, B1-4; Shanghai radio reports on *Wen Hui Pao* articles, 2 December 1968, in ibid., B4-5.
26. See pp. 61-62.
27. FBIS *China*, 17 March 1969, A1.
28. NCNA, 23 March 1969, in ibid., 24 March 1969, A1-3.
29. Richard M. Nixon, "Asia After Viet Nam," *Foreign Affairs*, October 1967, pp. 111-125.

CHAPTER SIX

1. For simplicity's sake, the Chinese name for the island will be used hereafter; this should not be taken as a judgment on the merits of either the Chinese or the Soviet arguments. The documentary film was reported by NCNA, 18 April 1969, in *PR* 12.17:3-4 (25 April 1969).
2. Col. Gen. O. Losik, "The Genealogy of Valor," *Izvestiya* morning ed., 3 December 1968, p.5, in FBIS *Soviet Union*, 16 December 1968, E1-2.
3. The article was broadcast in Vietnamese on 1 February. FBIS *China*, 5 February 1969, A4-5.
4. PRC Foreign Ministry Information Department document, "Chenpao Island Has Always Been Chinese Territory," in *PR* 12.11:14-15 (14 March 1969); NCNA's report on the documentary film cited in note 1; PRC Government Statement, 24 May 1969, in *PR* 12.22:4 (30 May 1969).

5. The note was broadcast as the 27th item on Moscow domestic radio's main evening newscast at 10:00 p.m. on 2 March, a quarter of an hour after TASS carried it in English. Text of the Soviet note in FBIS *Soviet Union*, 3 March 1969, A11-12.
6. *PR* 12.10:5,7 (7 March 1969).
7. Ibid., p. 12.
8. Ibid., pp. 6-7.
9. Statement by Soviet Foreign Ministry spokesman L. M. Zamyatin at 7 March 1969 press conference, in FBIS *Soviet Union*, 7 March 1969, A40-41.
10. *Pravda* editorial article, 8 March 1969, "A Provocative Sally of Peking Authorities," transmitted textually by TASS international service, in FBIS *Soviet Union*, 10 March 1969, A25-28.
11. *Red Star* editorial, 9 March 1969, in FBIS *Soviet Union*, 12 March 1969, A41-42.
12. See note 4 above.
13. *PR* 12.12:7 (21 March 1969).
14. Soviet Government Statement, 15 March 1969, in FBIS *Soviet Union*, 17 March 1969, A1-2.
15. NCNA account in *PR* 12.12:11-12 (21 March 1969). For the two sides' treatment of the territorial question at this time, see FBIS *Trends in Communist Propaganda*, 12 March 1969, pp. 11-13, and 19 March 1969, pp. 19-20.
16. FBIS *Soviet Union*, 1 April 1969, A1-7.

CHAPTER SEVEN

1. "On Summing Up Experience," *Red Flag* Nos. 3-4, 1969, in *PR* 12.12: 3-5 (21 March 1969).
2. *PR* 12.13:27 (28 March 1969).
3. The newly broadened claim was underscored by a curious orthographic innovation by NCNA's English service after the dissemination of a 14 April Congress communiqué announcing that the new constitution had been adopted. In a subsequent retransmission of the communiqué, NCNA introduced the new rendering "Mao Tsetung Thought"—removing the hyphen between the second and third elements of his name and dropping the possessive apostrophe. The effect was to give the visual impression, in English, that Mao Thought was on the same level and of the same kind of timeless authority as Marxism and Leninism. This could have been achieved less clumsily by using the term Maoist, but Peking never accepted that usage. For the 1969 CCP constitution see *PR* 12.18:36-39 (30 April 1969).

4. Ibid., pp. 16–35.
5. FBIS *Soviet Union*, 14 April 1969, A1.
6. *PR* 12.17:3–4 (25 April 1969).
7. For the TASS and *Pravda* reports on the Chinese film, see FBIS *Soviet Union*, 23 April 1969, A4. For the Kapitonov address, see ibid., B2–16, especially B15.
8. FBIS *Soviet Union*, 5 May 1969, A1.
9. *PR* 12.10:3 (16 May 1969).
10. Gromyko had cited border-river navigation as one of the subjects of Soviet proposals to the Chinese that year.
11. FBIS *Soviet Union*, 23 May 1969, A4.
12. *PR* 12.24:3, 39 (13 June 1969).

Chapter Eight

1. For an analysis of the 1957 and 1960 Moscow conferences, see Donald S. Zagoria, *The Sino-Soviet Conflict, 1956–1961* (Princeton, 1962), pp. 145–151, 343–369.
2. It should be noted that senior Soviet Politburo member Suslov's report to an important plenum in February 1964 included two major initiatives: the call for convening a new international party conference, and the disclosure of the expulsion of the "anti-party" group of leaders for being guilty of Stalinist crimes. Khrushchev was using both the China and the Stalin questions to undercut inhibiting forces in Kremlin politics. See Carl A. Linden, *Khrushchev and the Soviet Leadership, 1957–1964* (Baltimore, 1966), pp. 182, 184, 214.
3. FBIS *Soviet Union*, 18 June 1969, A21–47.
4. At the time of the 1960 Conference, Cuba had not yet been regarded as a Communist state.
5. "Internationalism Is the Banner of Communists," *Pravda*, 9 October 1968, pp. 4–5, in *Current Digest of the Soviet Press*, 30 October 1968, p. 7.
6. Nicola Shawi, "Internationalism and the Cause of Peace," *Pravda*, 23 October 1968, p. 4, in FBIS *Soviet Union*, 25 October 1968, A16–20.
7. Text of communiqué in ibid., 12 November 1968, A15–17.
8. Text of communiqué in ibid., 18 November 1968, A10. See FBIS *Survey*, 21 November 1968, pp. 19–21.
9. For developments at the Polish Congress and the East German Plenum, see FBIS *Survey*, 21 November 1968, pp. 21–23.
10. "To Strengthen the Cohesion of Communist Ranks," in FBIS *Soviet Union*, 26 November 1968, A21–24. See FBIS *Trends*, 27 November 1968, pp. 12–14.

11. FBIS *Survey*, 5 December 1968, pp. 21-23.
12. Ibid., 13 March 1969, pp. 3-4.
13. FBIS *Trends*, 12 March 1969, pp. 15-16.
14. Suslov's report was carried in abridged form by *Pravda*, 26 March 1969, pp. 1, 4, in FBIS *Soviet Union*, 28 March 1969, A6-12.
15. The condemnation of Stalin's adverse influence on the Comintern was the more notable in that it ran counter to the trend in the post-Khrushchev period of gradually rescuing Stalin's reputation from the denigration campaign promoted by Khrushchev. Thus, *Kommunist* earlier (No. 3, 1969) had repudiated a Khrushchev-era rehabilitation of a Stalin victim, Fedor Raskolnikov, who had escaped Stalin's purge by defecting to France in 1939 and had denounced the dictator. An article by five historians rebuked "some historians" for rehabilitating the likes of Raskolnikov, "who deserted to the camp of our enemies and slandered the party and Soviet state." See FBIS *Survey*, 13 March 1969, pp. 5-6.
16. *Pravda*, 26 March 1969, p. 4, in FBIS *Soviet Union*, 28 March 1969, A15-20.
17. "The Situation in China and the Position of the CCP at the Present Stage," *Kommunist* No. 4, 1969, pp. 86-104, in JPRS 48031, 13 May 1969, pp. 102-127.
18. "The Mao Tse-tung Policy in the International Arena," *Kommunist* No. 5, 1969, pp. 104-116, in JPRS 48107, 26 May 1969, pp. 121-135. For dissemination of Wang Ming's article, see FBIS *Trends*, 3 April 1969, pp. 13-14.
19. See *Scinteia*, 4 March 1969.
20. The Ivanov article was broadcast in Radio Moscow's Italian service; FBIS *Soviet Union*, 14 April 1969, A6-8. On the *L'Unita-Pravda* exchange, see FBIS *Trends*, 14 April 1969, pp. 9-11.
21. "The National Question in the Ideological Struggle," *Pravda*, 7 April 1969, pp. 2, 3, in FBIS *Soviet Union*, 9 April 1969, A13-19.
22. "To Strengthen the International Unity of Communists," *Kommunist* No. 7, 1969, pp. 3-11, in JPRS 48303, 26 June 1969, pp. 1-9.
23. O. Borisov and B. Koloskov, "The Anti-Soviet Course of the Mao Tse-tung Group," *Kommunist* No. 7, 1969, pp. 86-97, in JPRS 48303, 26 June 1969, pp. 79-92.
24. Final communiqué in FBIS *Soviet Union*, 2 June 1969, A42-43. For a discussion of the preparatory commission's meeting, see FBIS *Trends*, 4 June 1969, pp. 11-12.
25. Text of Suslov address in FBIS *East Europe*, 29 February 1968, RR16-24.
26. See note 3 above.
27. This proportion of the conference participants was about the same as

that of the delegations at the CPSU 22nd Congress in 1961 that condemned Albania. Reportage on the proceedings of the conference, including accounts of delegates' speeches, can be found in FBIS *Soviet Union* beginning on 6 June 1969.
28. TASS disseminated the text of Brezhnev's 7 June address worldwide. See FBIS *Soviet Union*, 9 June 1969, A37-67.
29. See pp. 137-138 and note 20 above.
30. See pp. 167-168.
31. See the TASS account of the 16 June proceedings in FBIS *Soviet Union*, 17 June 1969, A36-46.
32. See note 3 above.
33. For Suslov's Budapest meeting address, see note 25 above; for his 5 May 1968 speech on the Marx anniversary, see FBIS *Soviet Union*, 6 May 1968, B2-21.
34. "On the Results of the International Conference of Communist and Workers Parties," in ibid., 27 June 1969, B1-5.
35. "The Communist Movement Has Entered a Period of Upsurge," *Kommunist* No. 11, 1969, in JPRS 48818, 15 September 1969, pp. 1-16.
36. For further discussion of Brezhnev's article, see p. 185 on the border crisis.

CHAPTER NINE

1. *PR* 12.22:3-9 (30 May 1969).
2. Shortly after publication of the statement, a revolutionary committee—the governmental body established during the Cultural Revolution—was formed in the Ili Kazakh Autonomous Chou in Sinkiang, the site of the 1962 incident. Veteran Uighur official Saifudin, in a speech carried by the Sinkiang radio on 27 May 1969, mentioned the incident in the course of a revealing warning about continuing efforts by "national secessionists" with Soviet support to sever the Ili area from China. His reference to class enemies seeking to take advantage of "the religious issues of the national minorities" reflected the regime's concern over its vulnerabilities to Soviet troublemaking in this region. FBIS *China*, 29 May 1969, H1-3.
3. Ibid., 3 June 1969, A2-4.
4. See pp. 146-150 and note 28 of Chapter 8.
5. FBIS *Soviet Union*, 16 June 1969, A1-9.
6. J.G. Starke, *An Introduction to International Law* (London, 1963), pp. 173-174.
7. Col. Gen. P.I. Zyranov, "The Border Guard Is Always on the Alert," *Pravda*, 28 May 1969, p. 3, in FBIS *Soviet Union*, 2 June 1969, E5-6.
8. *PR* 12.24:4-5 (13 June 1969).

9. Ibid., p. 5.
10. FBIS *Soviet Union*, 12 June 1969, A1.
11. A. Mirov, "The Sinkiang Tragedy," *Literaturnaya Gazeta*, 7 May 1969, p. 15, in FBIS *Soviet Union*, 15 May 1969, A10-12.
12. *Pravda* 7 May 1969 article by A. Murzin, datelined Khorgos Village, Taldy-Kurgan Oblast alongside the Sinkiang border, excerpted in FBIS *Soviet Union*, 15 May 1969, A12-13.
13. *PR* 12.28:6 (11 July 1969).
14. FBIS *Soviet Union*, 9 July 1969, A1-2.
15. The Soviet note used the phrase "the Soviet part" of the island. Cf. a TASS Khabarovsk dispatch, 9 July, in FBIS *Soviet Union*, 9 July 1969, A2, and an NCNA account, 14 July 1969, in *PR* 12.29:21 (17 July 1969).
16. TASS dispatches in FBIS *Soviet Union*, 14 July 1969, A1-2.
17. A.P. Shitikov speech, *Izvestiya*, 13 July 1969, p. 3, in FBIS *Soviet Union*, 16 July 1969, B11-12.
18. FBIS *Soviet Union*, 11 August 1969, A1.
19. NCNA Harbin dispatch, 11 August, in FBIS *China*, 11 August 1969, A1.
20. "Strengthen Unity to Smash the Plot of Imperialism, Revisionism, and Reaction," *Red Flag* Nos. 6-7 (combined edition), 1969, in FBIS *China*, 16 July 1969, B1-4.
21. FBIS *China*, 5 August 1969, A2-5.
22. *New Times* (English edition), No. 27, 9 July 1969, pp. 11-12.
23. *PR* 12.33:3 (15 August 1969). See FBIS *Trends*, 13 August 1969, p. 14.
24. NCNA 13 August dispatch in *PR* 12.33:7 (15 August 1969).
25. FBIS *Soviet Union*, 13 August 1969, A20-21.
26. For an account of Soviet and Chinese propaganda in the aftermath of the 13 August clash, see FBIS *Trends*, 20 August 1969, pp. 9-15, and 27 August 1969, pp. 11-12. Amended 14 August NCNA dispatch in FBIS *China*, 15 August 1969, A1.
27. NCNA, 27 July 1969, in ibid., 28 July 1969, A1-2.
28. Ibid., 6 August 1969, A1-3.
29. *PR* 12.29:5-6 (18 July 1969).
30. FBIS *Soviet Union*, 11 July 1969, B1-23.
31. See note 35 of Chapter 8.
32. FBIS *Soviet Union*, 7 August 1969, E1-4.
33. Col. V. Klevtsov and Col. M. Novikov, "Khalkhin Gol, August 1939," *Pravda*, 19 August 1969, excerpted in FBIS *Soviet Union*, 21 August 1969, A25-26.
34. *Red Star* editorial, 19 August 1969, p. 1, in FBIS *Soviet Union*, 20 August 1969, A20-22.
35. *PR* 12.34:4-5 (22 August 1969).

36. TASS disseminated the 28 August *Pravda* editorial article worldwide. FBIS *Soviet Union*, 28 August 1969, A1-7.
37. Similarly, *Izvestiya* commentator V. Matveyev, speaking on a Moscow domestic radio panel discussion on 31 August, referred to "realistic-minded Americans" who understood the dangers of a Sino-Soviet war in the nuclear era. This remark appeared in a discussion in which he censured those in the United States who allegedly favored inciting the Chinese against the Soviet Union and in which he warned that a conflict of the magnitude of a Sino-Soviet war would be fraught with "almost unforeseeable consequences." FBIS *Soviet Union*, 2 September 1969, A70-72.
38. Ibid., 10 September 1969, A19-23.

CHAPTER TEN

1. Ho Chi Minh testament in FBIS *Asia & Pacific*, 10 September 1969, K5-7; Le Duan eulogy in ibid., 9 September 1969, K2-4.
2. Sino-Vietnamese communiqué in ibid., 11 September 1969, K1-2; Soviet-Vietnamese communiqué in ibid., 10 September 1969, K20-22.
3. Shanghai Writing Section of Revolutionary Criticism, "Who Are the Creators of History—Commenting on the Big Poisonous Weed 'The Beleaguered City,'" *Red Flag* No. 9, 1969, in FBIS *China*, 8 September 1969, B1-8.
4. As had happened a year earlier when the Peking leadership was debating new policy departures (see pp. 92-93), the issues raised in the *Red Flag* article figured in polemical attacks now resumed on the negotiationist approach again being considered. For example, an article in October invidiously compared "The Beleaguered City" with Chiang Ch'ing's model revolutionary work "Taking the Bandits' Stronghold." According to the article, the latter work sang the praise of people's war and was "a tribute to Comrade Lin Piao's implementation" of Mao's policy. On the other hand, "The Beleaguered City," which "promoted the capitulationist line" of Liu Shao-ch'i, had a hero who did not rely on people's war "but on the existing sharp dissensions within the enemy's ranks." As these polemical sallies demonstrate, a good source for apprehending another country's signals is the internal opposition's campaign against the policies being signaled. See the article broadcast by Peking domestic radio, 19 October 1969, "Splendid Epic of People's War and Art Specimen of Capitulationism," in FBIS *China*, 28 October 1969, B2-4.
5. In that prelude, the party's own leaders and cultural officials failed the test posed by Mao's demand at the 1962 Central Committee Plenum to practice class struggle and combat revisionism. Mandated by Lin Piao, Chiang Ch'ing conducted an exemplary cultural revolution within the army that presaged her major role in the later struggle to cleanse the party itself.

6. Moscow's original announcement of the Peking visit came more than 26 hours after the last previous report on his whereabouts, a TASS dispatch on 10 September announcing his arrival in Dushanbe, Tadzhikistan, on the "way back from Hanoi." TASS 10 September dispatch in FBIS *Soviet Union*, 11 September 1969, A26.
7. Ibid., A26.
8. FBIS *China*, 12 September 1969, A1. Kosygin was accompanied by K. Katushev, the Central Committee secretary in charge of relations with governing parties, and a Supreme Soviet official. Chou's associates at the meeting were his Deputy Premier, Li Hsien-nien (also just back from Hanoi), and Hsieh Fu-chih, who perhaps was present as Mayor of Peking. The Chinese report cited only Kosygin's position of Premier, not his party title, and simply referred to Katushev by his name, thus avoiding any acknowledgment of party-level relations. By contrast, a Romanian delegation stopping over on the same day was met by K'ang Sheng, who handled inter-party relations, as well as by Chou and Li, and the Peking announcement cited party titles and said the two sides had a "friendly" conversation. Ibid., A1.
9. In addition to avoiding new polemics, Moscow's stand-down was reflected in the omission of any reference to anti-Chinese articles from press reviews of issues of journals already in the publication pipeline before the Kosygin-Chou meeting. For Soviet propaganda treatment of China in the wake of the meeting, see FBIS *Trends*, 17 September 1969, pp. 12–14, and 24 September, pp. 11–13.
10. FBIS *Soviet Union*, 24 September 1969, A1.
11. K. Rodionov, "Soviet Foreign Policy Documents," *Sovetskaya Rossiya*, 19 September 1969, p. 3, in ibid., A51–52.
12. In addition to *Pravda*, Radio Moscow's foreign services played up the Vietnamese editorials appealing for international Communist unity. See the Glazunov commentaries in FBIS *Soviet Union*, 17 September 1969, A4–6, and 22 September, A13–14.
13. For the attribution to Mao, see the joint editorial on New Year's Day, 1970, in *PR* 13.1:7 (2 January 1970). National Day slogans in *PR* 12.38: 3–4 (19 September 1969).
14. NCNA report on the Korean delegation's arrival in FBIS *China*, 1 October 1969, A9–10. The slogan on a possible war launched by "imperialism or social-imperialism" was used for the Korean delegation as for others. Pyongyang's concern over the effects on the Communist alliance system, and hence on North Korea's security, of an uncontrolled Sino-Soviet conflict had been in evidence since the escalation of the Vietnam War in 1965. This concern had recently been registered in a lengthy *Nodong Sinmun* article, disseminated internationally in two installments by

KCNA on 27-28 August 1969, which pleaded for unity against the common enemy, the United States, and obliquely rebuked both Moscow and Peking: "It is wrong either to trample independence underfoot under the signboard of internationalism or to betray internationalism under the cloak of independence in the international communist movement." The first half of that rebuke (addressed to Moscow) throws some light on the peculiar way in which Pyongyang the previous year had stressed the dangers of the Czechoslovakia liberalization while barely mentioning the actual Soviet intervention there. The North Koreans were all in favor of squelching that liberalization—in their eyes an acute form of the revisionist malady—but the Soviet invasion could easily be seen as "trampling independence underfoot under the signboard of internationalism." For the August 1969 *Nodong Sinmun* article, see FBIS *Asia & Pacific*, 28 August 1969, D1-6.
15. *People's Daily/Red Flag/Liberation Army Daily* joint editorial, 1 October 1969, in *PR* 12.40:19-21 (3 October 1969).
16. Chou En-lai speech at National Day reception in ibid., pp. 17-18; Chou speech at 27 September dinner for the Cambodian delegation in FBIS *China*, 29 September 1969, A18-20.
17. Ibid., 6 October 1969, B1.
18. Chu Kuang-ya, "Rely on Mao Tse-tung Thought To Scale the Heights of Science and Technology," *Red Flag* No. 10, 1969, in ibid., B9-13.
19. *PR* 12.40:15-16 (3 October 1969).
20. FBIS *Soviet Union*, 1 October 1969, A1.

Chapter Eleven

1. *PR* 12.41:3-4 (10 October 1969).
2. Ibid., pp. 8-15.
3. FBIS *China*, 20 October 1969, A1.
4. FBIS *Soviet Union*, 20 October 1969, A1.
5. FBIS *China*, 24 October 1969, A4-7.
6. FBIS *Soviet Union*, 27 October 1969, A20.
7. Ibid., 28 October 1969, A1-10.
8. A watershed *Pravda* editorial article on 27 November 1966 introducing the anathematizing formula "Mao Tse-tung and his group" had signaled the departure from the protocol of addressing the Chinese as "comrades." The "Mao group" formula signified that, in Moscow's eyes, a Chinese faction had illegitimately usurped the authority of a Communist party.
9. Podgornyy 6 November address in FBIS *Soviet Union*, 7 November 1969, B1-16.

10. FBIS *China*, 9 October 1969, A14-16.
11. Pham Van Dong speech cited in note 5. Sino-Vietnamese communiqué in ibid., 27 October 1969, A1-3.
12. See pp. 197-199.
13. These trailblazing advances were made by the province headed by Hua Kuo-feng, who was destined for higher levels.
14. "It Is Necessary to Firmly Grasp the 'Four Good,'" in FBIS *China*, 15 October 1969, B1-6.
15. *PR* 13.1:5-7 (2 January 1970).
16. *Ta Kung Pao*, Hong Kong, English edition, 6-12 November 1969, in FBIS *China*, 6 November 1969, A1.
17. Divergent Chinese and Soviet reports in ibid., 15 December 1969, A1.
18. Ibid.
19. Peking's announcement in ibid., 12 January 1970, A1; for McCloskey's reference to the PRC, see Tad Szulc, *New York Times*, 15 February 1970, p. 6.
20. FBIS *China*, 9 January 1970, A1.
21. FBIS *Soviet Union*, 9 January 1970, A1.
22. See pp. 197-199.
23. S. Tikhvinskiy, "Geopolitical Fortunetelling," *Pravda*, 15 February 1970, pp. 4-5, in FBIS *Soviet Union*, 17 February 1970, A8-13.
24. Szulc, *New York Times*, 15 February 1970, p. 6.

CHAPTER TWELVE

1. See pp. 229-230.
2. *People's Daily* editorial, 5 May 1970, "The New War Adventure of the Nixon Administration," in FBIS *China*, 5 May 1970, A2-3. Documents on the Indochinese summit conference sponsored by the Chinese in *PR* 13, Special Edition (8 May 1970) devoted to the event.
3. FBIS *China*, 19 May 1970, A1.
4. Ibid., 20 May 1970, A1-2.
5. Ibid., 20 June 1970, A1.
6. Ibid., 10 July 1970, A1.
7. Ibid., 30 June 1970, B1-3; also in *PR* 13.27:10-11 (3 July 1970).
8. Worker-PLA Mao Tse-tung Thought Propaganda Team Stationed at Peking University, "Struggle To Strengthen the Building of the Party Ideologically," *Red Flag* No. 7, 1970, in FBIS *China*, 9 July 1970, B1-11.
9. CCP Committee of the Shenyang Metallurgical Plant, "Strengthen the Building of the Party Committee Ideologically," *Red Flag* No. 7, 1970, in ibid., 15 July 1970, B1-8.

10. Text of communiqué in *PR* 13.37:5-7 (11 September 1970).
11. See FBIS *Trends*, 10 September 1970, pp. 36-37, and 9 December 1970, p. 34.
12. Marvin Kalb and Bernard Kalb, *Kissinger* (New York, 1975), p. 174.
13. FBIS *China*, 13 October 1970, A14.
14. *People's Daily* editorial, 15 October 1970, "Welcome the Establishment of Diplomatic Relations Between China and Canada," in ibid., 15 October 1970, A1-2.
15. Ibid., 1 December 1970, A14-16.
16. Documents on the Chinese celebration of the anniversary in *PR* 13.44: 5-25 (30 October 1970); Huang's speech at pp. 9-11.
17. *New York Times*, 11 December 1970, p. 32 (emphasis added).
18. Richard M. Nixon, *U. S. Foreign Policy for the 1970's—Building for Peace*, Report to Congress, 25 February 1971 (Washington, D.C., 1971); NCNA account in FBIS *China*, 5 March 1971, A1-4.
19. Edgar Snow, "A Conversation with Mao Tse-tung," *Life*, 30 April 1970, pp. 46-48.
20. *PR*, 14.8:6 (19 February 1971).
21. *People's Daily* editorial, 14 February 1970, "All-out Support to Peoples of Three Countries in Indochina in War Against U. S. Aggression and for National Salvation," ibid., p. 7.
22. *People's Daily* "Commentator," "Don't Lose Your Head, Nixon," 20 February 1971, in *PR* 14.9:6 (26 February 1971).
23. Chou En-lai speech in FBIS *China*, 10 March 1971, A9-14; Pham Van Dong speech in ibid., 11 March 1971, A8-10.
24. Ibid., 10 March 1971, A14-20 (emphasis added).
25. Ibid., 17 March 1971, A6-8.
26. Ibid., 23 March 1971, A7-10.
27. Ibid., 15 April 1971, A1-3.
28. Shanghai broadcast in ibid., 16 April 1971, A1-2; report on Sihanouk banquet in ibid., 14 April 1971, A1-2.
29. PLA Chief of Staff Huang Yung-sheng speech at 27 June 1970 Pyongyang rally, in *PR* 13.27:54-55 (3 July 1970).
30. For the Lodge report, see *New York Times*, 27 April 1971, p. 1; for Nixon press conference, see ibid., 30 April 1971, p. 18.
31. *People's Daily* "Commentator," 4 May 1971, in FBIS *China*, 5 May 1971, A37-38; for Bray's remarks, see *New York Times*, 29 April 1971, p. 4.
32. Chou speech in FBIS *China*, 9 June 1971, A17-19; Sino-Romanian communiqué in ibid., A19-24.
33. See pp. 90-95.
34. *People's Daily/Red Flag/Liberation Army Daily* joint editorial, 1 Jan-

uary 1971, "Advance Victoriously Along Chairman Mao's Revolutionary Line," in *PR* 14.1:8-10 (1 January 1971).
35. FBIS *China*, 11 January 1971, B1-6.
36. "Victory for Chairman Mao's Line on Party Building," in ibid., 1 February 1971, B2-4.
37. The Revolutionary Committee of the Anch'ing District, Anhwei Province, "Conscientiously Sum Up Experience and Overcome Complacency," *Red Flag* No. 2, 1971, in ibid., 12 February 1971, B1-6.
38. The Party Committee of a Division Under the PLA Tsinan Units, "Remain Modest and Prudent, Continue To Make Revolution," *Red Flag* No. 2, 1971, in ibid., 23 February 1971, B1-7.
39. The Writing Group of the Kiangsi Provincial CCP Committee, "Uphold the Materialist Theory of Reflection, Criticize Idealist Transcendentalism, and Study 'Where Do Correct Ideas Come From?'" *Red Flag* No. 4, 1971, reprinted in the 11 April editions of *People's Daily* and *Liberation Army Daily* and broadcast that day by the Peking domestic radio, in FBIS *China*, 13 April 1971, B1-9.
40. *PR* 14.19:10-12 (7 May 1971).
41. *PR* 14.32:7-9 (6 August 1971). See FBIS *Trends*, 4 August 1971, pp. 23-24.
42. See note 41 above.
43. The Writing Group of the Hupeh Provincial CCP Committee, "Strong Weapon to Unite the People and Defeat the Enemy—Study 'On Policy,' " *Red Flag* No. 9, 1971, in FBIS *China*, 18 August 1971, B1-7.
44. See note 18 of Chapter 3.
45. *PR* 15.9:4-5 (3 March 1972).
46. Text of Brezhnev 20 March 1972 speech in FBIS *Soviet Union*, 21 March 1972, J1-17. See FBIS *Trends*, 22 March 1972, pp. 8-11.
47. *New York Times*, 9 May 1972, p. 18.
48. Ibid., pp. 1, 18.
49. FBIS *Soviet Union*, 30 May 1972, AA3-5.
50. *Nhan Dan* editorial, 17 August 1972, "Victory of the Revolutionary Trend," in FBIS *Asia & Pacific*, 17 August 1972, K5-9.
51. *Nhan Dan* 19 August editorial marking the anniversary of the August 1945 Communist revolution, in ibid., 21 August 1972, K1-3. See FBIS *Trends*, 23 August 1972, pp. 1-6.
52. See note 50 above.
53. FBIS *Asia & Pacific*, 11 September 1972, L1-6. See FBIS *Trends*, 13 September 1972, pp. 1-6.
54. *Hoc Tap*, "Two Opposing Negotiating Stands at the Paris Talks," September 1972, in FBIS *Asia & Pacific*, 28 September 1972, K1-7. See FBIS *Trends*, 27 September 1972, pp. 1-2.

Chapter Thirteen

1. Neville Maxwell, "The Chinese Account of the 1969 Fighting at Chenpao," *China Quarterly* 56 (October/December 1973) pp. 730–739.

Bibliography

Austin, J.L. *How To Do Things With Words.* New York, Oxford University Press, 1970.

Bridgham, Philip L. "The International Impact of Maoist Ideology," in Chalmers Johnson, ed., *Ideology and Politics in Contemporary China.* Seattle, University of Washington Press, 1973.

Clubb, O. Edmund. *China and Russia: The "Great Game."* New York and London, Columbia University Press, 1971.

Current Digest of the Soviet Press

Doolin, Dennis J. *Territorial Claims in the Sino-Soviet Conflict.* Stanford, Hoover Institution, Stanford University, 1965.

Foreign Broadcast Information Service. *Daily Report: Asia & Pacific, Communist China* (from August 1971, *People's Republic of China*), *East Europe, Latin America, Soviet Union.* Springfield, Virginia, National Technical Information Service, U.S. Department of Commerce.

———. *Survey of Communist Propaganda.* Springfield, Virginia, National Technical Information Service, U. S. Department of Commerce.

———. *Trends in Communist Propaganda.* Springfield, Virginia, National Technical Information Service, U.S. Department of Commerce.

Franck, Thomas M. and Edward Weisband. *Word Politics.* New York, Oxford University Press, 1972.

Garthoff, Raymond L., ed. *Sino-Soviet Military Relations.* New York, Frederick A. Praeger, 1966.

Griffith, William E. *The Sino-Soviet Rift.* Cambridge, The M.I.T. Press, 1964.

———. *Sino-Soviet Relations, 1964–1965.* Cambridge, The M.I.T. Press, 1967.

Joint Publications Research Service. Springfield, Virginia, National Technical Information Service, U.S. Department of Commerce.

Kalb, Marvin and Bernard Kalb. *Kissinger.* New York, Dell, 1975.

Karnow, Stanley. *Mao and China*. New York, The Viking Press, 1972.

Linden, Carl A. *Khrushchev and the Soviet Leadership, 1957–1964*. Baltimore, The Johns Hopkins Press, 1966.

Mao Tse-tung. "On Policy," in *Selected Works* (English edition), Vol. 2. Peking, Foreign Languages Press, 1965.

Maxwell, Neville. *India's China War*. Garden City, N.Y., Anchor Books, 1972.

——. "The Chinese Account of the Fighting at Chenpao," in *China Quarterly*, 56 (October/December 1973), pp. 730–739.

New Times, Moscow (English edition of *Novoye Vremya*).

New York Times.

Nixon, Richard M. "Asia After Viet Nam," *Foreign Affairs*, October 1967, pp. 111–125.

——. *United States Foreign Policy for the 1970's: Building for Peace*. Washington, D.C., Government Printing Office, 1971.

Peking Review (weekly).

Salisbury, Harrison E. *War Between Russia and China*. New York, W.W. Norton and Company, 1969.

Snow, Edgar. *Red Star Over China*. New York, Random House Modern Library, 1944.

——. "A Conversation With Mao Tse-tung," *Life*, 30 April 1970, pp. 46–48.

Starke, J.G. *An Introduction to International Law*. London, Butterworths, 1963.

Szulc, Tad. "Behind the Vietnam Cease-fire Agreement," *Foreign Policy*, 15 (Summer 1974), pp. 21–69.

Ta Kung Pao, Hong Kong (Chinese and English editions).

Wich, Richard. "Chinese Allies and Adversaries," in William W. Whitson, ed., *The Military and Political Power in China in the 1970s*. New York, Praeger Publishers, 1972.

Winch, Peter. *The Idea of a Social Science and its Relation to Philosophy*. New York, Humanities Press, 1958.

Zagoria, Donald S. *The Sino-Soviet Conflict, 1956–1961*. Princeton, Princeton University Press, 1962.

Index

Albania, 4, 45, 55, 67-71, 155, 218, 273
Amur River, 27, 101, 106, 119, 120, 169, 171, 173, 211
Anti-ballistic missile, 94
Asian collective security system, 168, 181, 182-183
Austin, J.L., 19

Bakdash, Khalid, 128
Balluku, Beqir, 69, 70, 73
Bandung Conference, 10
Berlin, 8, 97
Berlinguer, Enrico, 151-152
Bilak, Vasil, 140
Bipolarity, 8, 9-10, 11-13, 15-16, 158, 210, 218, 231, 270, 277-278, 281-282
Boffa, Giuseppe, 137-138
Bragamov, E., 138
Brandt, Willy, 215, 216
Bray, Charles, 250-251
Brezhnev, L.I., 17, 22-23, 107, 130, 139, 158, 160-161, 164, 181, 182, 185, 213-215, 223, 229, 262-263, 275; doctrine of limited sovereignty, 15, 130, 138, 147, 276; speech at Moscow International Conference (1969), 141, 146-150, 156, 167-168, 189
Bulganin, N.A., 26
Bulgaria, 69, 142-143

Cairo declaration, 250
Cambodia, 232-235, 248. *See also* Sihanouk
Canada, 239-240, 261
Carillo, Santiago, 137

Castro, Fidel, 4, 46, 47-49, 51
Ceausescu, Nicolae, 46, 131, 142, 147, 150-151, 152, 251
Chen I, 63-64, 85
Ch'en Po-ta, 236-238, 252
Chenpao (Damanskiy) Island, 98, 102, 106, 107, 108, 111, 117, 120, 164-165, 166, 168, 176, 187, 211, 212, 274-275
Chi Peng-fei, 103
Chiang Ch'ing, 198-199, 228, 229
Chiang Kai-shek. *See* Chinese Nationalists
Ch'iao Kuan-hua, 217, 240
China, People's Republic of: relations with Japan, 3, 89, 240, 268, 269; relations with Vietnam, 3, 7, 59-63, 80-81, 82, 84, 196-198, 218-220, 234-235, 239, 245-247; East Europe policy, 3-4, 5, 56, 58, 67, 73, 158, 218, 271, 273; East Asia policy, 5, 6; internal politics, 17-18, 75-76, 85-87, 92-93, 199, 220-223, 235-238, 251-256, 271-273; relations with Mongolia, 26-27; Cultural Revolution, 44, 72, 75-80, 86, 87, 92-93, 95, 103, 114, 115-116, 123, 198, 199, 205, 213, 221, 223, 235-238, 252-256, 270-273; reaction to invasion of Czechoslovakia, 55-64; party rebuilding, 77, 199, 221, 235-237, 252-256; and CCP Ninth Congress, 77, 79, 95, 104, 114, 114-118, 123, 148, 163, 167, 202, 248, 252, 255, 272, 275; U.N. representation, 82, 88-89, 242-244, 249-250; and Soviet-U.S. relations, 89-90, 93-94;

China, People's Republic of: *(continued)* National Day (1969) slogans, 202-204; nuclear tests, 205; National People's Congress, 237, 253; General Political Department of PLA, 238; relations with North Korea, 241-242; reaction to Lam Son 719 operation (1971), 244-247. *See also* Sino-American relations; Sino-Soviet border; Sino-Soviet relations
Chinese Civil War, 8, 9, 15, 43, 87. *See also* Taiwan
Chinese Nationalists, 9, 10, 11, 87, 181, 201, 243-244, 250, 252
Chou En-lai, 26, 33-34, 69, 86, 87, 93, 100-102, 117, 136, 142, 181, 183, 203, 214, 218, 219, 223, 229, 233, 241, 248, 251, 272; speech on Romanian National Day (1968), 56-57, 61-62, 68; speech on DRV National Day (1968), 57, 59-62, 69, 100, 193; at Ho Chi Minh funeral, 196-197, 198; meeting with Kosygin, 199-200, 202, 204, 208, 213, 227-228, 272, 275; visit to North Korea (1970), 204; PRC National Day address (1969), 204-205; head of delegation to Hanoi (1971), 246-247
Cominform, 124, 125, 127, 132
Comintern, 124, 125, 131-134
Communist Party of Soviet Union, 1, 11, 15, 27, 63, 124, 158, 270. *See also* Soviet Union
Communist Party of USA, 28, 144
Content analysis, 20, 280
Contextual analysis, 20, 280-283
Costa Rica-Nicaragua Treaty, 169, 211
Cuba, 4, 28, 47-49, 63, 152-153
Czechoslovakia, 2, 4, 7, 15-16, 38, 41-42, 66, 98-100, 183, 188-189, 214-215, 268, 269, 270, 272, 273, 276, 277, 279, 282, 283; and Moscow international conference (1969), 127-128, 139-140, 144-146

Damanskiy Island. *See* Chenpao
Dimitrov, Georgi, 142
Dobrynin, A.F., 216
Dong, Pham Van, 53-54, 203, 213, 219-220, 246

Duan, Le, 196
Dubcek, Alexander, 58, 139, 214
Dulles, John Foster, 182

Eisenhower, Dwight D., 12
Equatorial Guinea, 240
Esoteric communication, 20, 280, 282

France, Communist Party of, 47, 129, 137-138

Gomulka, Wladyslaw, 129, 142
Gromyko, A.A., 36-37, 121, 182-184, 185, 223
Guevara, Ernesto (Che), 4

Hager, Kurt, 130
Haiphong harbor, 232, 264-265
Hall, Gus, 144
Henry, Ernst, 191
Ho Chi Minh, 193-194, 196-197, 198, 201, 220, 275
Hong Kong, 28
Huang Yung-sheng, 70-71, 73, 242, 245
Hungary, 11, 63, 144, 145, 215, 231, 278
Husak, Gustav, 139, 146, 214

Ilichev, L.F., 263
India, 12, 37-39, 42, 108
Ionov, Lt. Gen. P., 111
Israel, 155
Italy, Communist Party of, 47, 129, 130, 148, 151-152
Ivanov, I., 137-138

Japan, 3, 89, 240, 268, 269
Johnson, Lyndon B., 6, 22, 44, 58, 68, 81, 91, 227

Kadar, Janos, 144, 145
Kapitonov, I.V., 119
Kazakhs, 30, 36, 144
Khalkhin Gol, 187
Khrushchev, N.S., 11, 12, 17, 26, 27, 28, 31-32, 33-34, 36-37, 63, 106, 112, 123, 124, 126, 142, 146, 149, 155, 164, 177, 194, 273, 278; policy on China, 17, 26, 27, 28, 31-32, 33-34, 36-37, 106, 177; and international

Khrushchev, N.S., *(continued)*
  Communist conferences, 37, 112, 123, 126, 146, 155, 273
Kim Il-song, 203
Kirilenko, A.P., 129
Kissinger, Henry, 88, 230, 232, 238, 256, 257, 264-265, 266-267
Kommunist editorial articles on China, 134-135, 136-137; editorial on 1969 International Party Conference, 138-139
Korea, 8-9, 10, 15, 49-51, 63, 194, 203, 241-242, 245, 246, 249, 278
Kosygin, A.N., 107, 117, 194, 196, 274; meeting with Chou En-lai, 185, 199-200, 202, 204, 208, 213, 227-228, 272, 275
Kuznetsov, V.V., 212, 225

Laos, 232, 244-247
Lei Yang, 226, 227
Li Hsien-nien, 196, 240
Li Te-sheng, 238
Lin Piao, 60, 68, 75-76, 78, 93, 95, 101, 198-199, 203, 205, 222, 229, 238, 252, 256, 259-260, 272, 273; on people's war, 60, 79, 86, 198, 259, 277, 281; and Mao's Thought, 78, 86; as Mao's successor designate, 114, 123; Ninth Congress political report, 114-118, 202
Liu Shao-ch'i, 77, 92, 115-116, 236, 255
Lodge, Henry Cabot, 220, 249
Longo, Luigi, 47

McCloskey, Robert, 227, 234
Macao, 28
Mao Tse-tung, 25-26, 34-35, 68, 86-87, 90, 93, 100, 104, 113, 115-116, 123, 135, 149, 166, 199, 201, 202, 203, 205, 231, 236, 244, 250, 252, 256, 260; interview with Japanese socialists (1964), 34-35, 144; Thought of, 77-80, 86, 114, 115, 256; inscription for Japanese workers (1962), 78-79, 272; 20 May 1970 statement, 234-235; "On Policy," 257-260
Matrosov, V.A., 225
Mongolia, 25-26, 45, 143-144, 178

Moscow International Communist Conference of 1957, 124, 125, 141, 149, 152, 157
Moscow International Communist Conference of 1960, 124-125, 127, 132, 139, 141, 152, 155, 156-157
Moscow International Communist Conference of 1969, 23, 41-42, 55-56, 112, 123, 126-127, 163, 167, 171-172, 183, 275; preparations for, 127-140; proceedings of, 140-153; main document, 141, 151, 153-158, 159; treatment of China, 141-144, 155, 160; treatment of Czechoslovakia, 144-146, 154-155; treatment of nationalism, 156-157, 159

Nixon, Richard M., 2, 5, 81-82, 87-90, 91, 92, 94, 136, 184, 216, 219-220, 223, 226-227, 233-234, 238-239, 245-246, 264-265; visit to China, 2, 230, 232, 256, 257, 259, 260-263, 268; visit to USSR, 262, 264-266; Nixon Doctrine, 180-181, 218, 241, 243-244, 248, 249-250, 251

Ochetto, Achille, 130

Peaceful coexistence, 10, 14, 83, 84-85, 149, 156, 205, 209-210, 217, 223, 240-241, 263, 264, 279
P'eng Te-huai, 197
Phillips, Christopher, 243
Podgornyy, N.V., 216, 266
Ponomarev, B.N., 133-134, 153-154
Potsdam declaration, 250

Red Flag: on "The Beleaguered City," 197-199, 220, 228, 272-273; on Mao's "On Policy," 257-260
Redmond, Hugh Francis, 235
Rochet, Waldeck, 47
Rodriguez, Carlos Rafael, 152-153
Romania, 46, 131-132, 137, 138, 142, 144, 147, 150-151, 154, 155, 157-158, 251; relations with China, 3, 57, 71, 274

Saifuddin, 76
Salisbury, Harrison, 229-230

Index

Sato, Eisaku, 89, 240
Shawi, Nicola, 295
Shehu, Mehmet, 67–68
Sihanouk, Norodom, 205, 232–233, 247, 248
Sinkiang, 26, 27, 29, 30–31, 33, 36, 37, 71, 76, 171–173, 176–179
Sino-American relations, 2–3, 5, 9, 18, 58, 81–85, 87–92, 93–95, 180–181, 227, 231–232, 233–234, 235, 242, 243–244, 248–252, 269, 271, 278–280; ambassadorial talks, 10, 83–85, 87, 88–89, 91, 226–227, 229, 230, 233–235; Shanghai communiqué, 58, 259, 260–263, 264. *See also* China, People's Republic of; Nixon; Taiwan
Sino-Soviet border, 5, 7, 16, 93–94, 164, 184–192, 229–230, 266, 268, 270–276; troubles on, 27–28, 29–33, 38, 65–67, 69, 71–72, 97–98, 101–103, 109–110, 171–175, 176–180, 188, 191–192; treaties on, 28–29, 35–36, 38–39, 102, 103, 106–108, 110–111, 117, 163–164, 165–166, 168–170, 174, 207, 210–211; talks on, 32, 35, 40, 106–108, 111–112, 117, 118, 119, 139, 164–171, 182, 189, 191, 197, 200–201, 204, 206, 207–208, 209–214, 216, 221, 224–226, 227–228, 263, 269, 274–276; Chinese charges of Soviet troop concentrations, 69–70, 99, 103, 107, 167, 177, 179; Chinese charges of Soviet invasion threat, 70–71, 202, 208–209; thalweg principle, 106, 166, 168–169, 173–174, 208, 211–212; joint river navigation commission on, 119–122, 174–176, 182, 276; *Pravda* editorial article (August 1969) on, 189–191, 193, 194, 197, 204. *See also* Amur; Chenpao; Mongolia; Sinkiang; Ussuri
Sino-Soviet relations, 1, 2, 9–10, 11, 12, 13, 14–16, 43, 261, 263, 269–270, 276–278; treaty of alliance, 10, 12, 85, 229, 231, 268. *See also* Sino-Soviet border
Snow, Edgar, 244, 250
Socialist camp, 2, 6, 7, 9, 12–13, 15, 16, 45, 48, 51, 52–54, 59–61, 63, 69, 75, 79, 100, 127, 136, 195, 196, 202, 210, 231, 269, 277

Socialist community, 15, 56, 60–61, 63
Southeast Asia Treaty Organization (SEATO), 10, 181
Soviet Union, 1, 4, 10, 11, 125; as China's primary adversary, 2, 16, 61–62, 89–90, 182, 231–232, 258–260, 269; as "social imperialism," 7, 56, 57, 59, 62, 75, 99, 116, 203, 204, 274; united action policy on Vietnam, 14, 45, 54, 123, 126, 194, 195, 259, 271, 273; relations with United States, 149–150, 183–184, 264–265. *See also* Moscow International Communist Conferences; Sino-Soviet border; Sino-Soviet relations
Stalin, I.V., 1, 11, 26, 124, 132, 133, 134, 231
Stoessel, Walter, 226, 227
Strategic Arms Limitation Talks (SALT), 41–42, 89–90, 183, 215–216, 225, 234, 264
Suslov, M.A., 133, 140–141, 148, 157

Taipov, Zunun, 30, 177–178
Taiwan, 8, 9, 10, 11, 12, 26, 28, 34, 42, 83–85, 88, 181, 226, 233, 239–240, 241, 243–244, 248–251, 252, 261–262, 278–280; U.S. "occupation" of, 9, 33, 88, 223, 242, 249
Thieu, Nguyen Van, 252, 267
Tho, Le Duc, 266–267
Thuy, Xuan, 238
Tibet, 26, 37, 38, 76, 178
Tikhvinskiy, S., 229–230
Tito, Josip Broz, 5, 7, 132
Togliatti, Palmiro, 151
Tolubko, V.P., 186–187
Tsedenbal, Yumjaagiin, 143–144
Tung Pi-wu, 240

Uighurs, 30, 76, 144, 173, 177–178
United States, 1, 8, 9, 18, 22, 28, 41, 48, 49, 51, 66–67, 68, 136, 146, 149, 180, 202, 218, 233, 258; and Soviet Union, 11–12, 22, 43–44, 54, 55–56, 57, 58–59, 61, 63–64, 89–90, 93–94, 97, 182–184, 215–216, 223. *See also* Nixon; Sino-American relations; Soviet Union; Taiwan; Vietnam
Ussuri River, 27, 97, 101, 106, 107, 111, 117, 119, 120, 168–169, 171, 211, 274

Vietnam, 2, 5, 6, 8, 9, 10, 14, 15, 16, 43–44, 52, 80–81, 84, 85, 154, 180, 183–184, 194–197, 218–220, 232, 238–239, 244–248, 259, 263–268, 277, 281; talks on, 44, 80–81, 84, 85, 220, 238–239, 265–268. *See also* China, People's Republic of; Vietnam, Democratic Republic of

Vietnam, Democratic Republic of, 6, 7, 45, 51–55, 193–196, 218–220, 265–268. *See also* China, People's Republic of; Vietnam

Walsh, Bishop James Edward, 235

Wang En-mao, 71
Wang Ming, 135–136

Yakubovskiy, I.I., 186–187
Yao Wen-yuan, 93, 116
Yeh Chien-ying, 70
Yugoslavia, 3–4, 5, 10, 45, 46, 124, 132, 150, 218, 240, 274

Zamyatin, L.M., 111, 200–201
Zhivkov, Todor, 142–143
Zyryanov, P.I., 171

Harvard East Asian Monographs

1. Liang Fang-chung, *The Single-Whip Method of Taxation in China*
2. Harold C. Hinton, *The Grain Tribute System of China, 1845–1911*
3. Ellsworth C. Carlson, *The Kaiping Mines, 1877–1912*
4. Chao Kuo-chün, *Agrarian Policies of Mainland China: A Documentary Study, 1949–1956*
5. Edgar Snow, *Random Notes on Red China, 1936–1945*
6. Edwin George Beal, Jr., *The Origin of Likin, 1835–1864*
7. Chao Kuo-chün, *Economic Planning and Organization in Mainland China: A Documentary Study, 1949–1957*
8. John K. Fairbank, *Ch'ing Documents: An Introductory Syllabus*
9. Helen Yin and Yi-chang Yin, *Economic Statistics of Mainland China, 1949–1957*
10. Wolfgang Franke, *The Reform and Abolition of the Traditional Chinese Examination System*
11. Albert Feuerwerker and S. Cheng, *Chinese Communist Studies of Modern Chinese History*
12. C. John Stanley, *Late Ch'ing Finance: Hu Kuang-yung as an Innovator*
13. S. M. Meng, *The Tsungli Yamen: Its Organization and Functions*
14. Ssu-yü Teng, *Historiography of the Taiping Rebellion*
15. Chun-Jo Liu, *Controversies in Modern Chinese Intellectual History: An Analytic Bibliography of Periodical Articles, Mainly of the May Fourth and Post-May Fourth Era*
16. Edward J. M. Rhoads, *The Chinese Red Army, 1927–1963: An Annotated Bibliography*
17. Andrew J. Nathan, *A History of the China International Famine Relief Commission*
18. Frank H. H. King (ed.) and Prescott Clarke, *A Research Guide to China-Coast Newspapers, 1822–1911*
19. Ellis Joffe, *Party and Army: Professionalism and Political Control in the Chinese Officer Corps, 1949–1964*
20. Toshio G. Tsukahira, *Feudal Control in Tokugawa Japan: The Sankin Kōtai System*

21. Kwang-Ching Liu, ed., *American Missionaries in China: Papers from Harvard Seminars*
22. George Moseley, *A Sino-Soviet Cultural Frontier: The Ili Kazakh Autonomous Chou*
23. Carl F. Nathan, *Plague Prevention and Politics in Manchuria, 1910–1931*
24. Adrian Arthur Bennett, *John Fryer: The Introduction of Western Science and Technology into Nineteenth-Century China*
25. Donald J. Friedman, *The Road from Isolation: The Campaign of the American Committee for Non-Participation in Japanese Aggression, 1938–1941*
26. Edward Le Fevour, *Western Enterprise in Late Ch'ing China: A Selective Survey of Jardine, Matheson and Company's Operations, 1842–1895*
27. Charles Neuhauser, *Third World Politics: China and the Afro-Asian People's Solidarity Organization, 1957–1967*
28. Kungtu C. Sun, assisted by Ralph W. Huenemann, *The Economic Development of Manchuria in the First Half of the Twentieth Century*
29. Shahid Javed Burki, *A Study of Chinese Communes, 1965*
30. John Carter Vincent, *The Extraterritorial System in China: Final Phase*
31. Madeleine Chi, *China Diplomacy, 1914–1918*
32. Clifton Jackson Phillips, *Protestant America and the Pagan World: The First Half Century of the American Board of Commissioners for Foreign Missions, 1810–1860*
33. James Pusey, *Wu Han: Attacking the Present through the Past*
34. Ying-wan Cheng, *Postal Communication in China and Its Modernization, 1860–1896*
35. Tuvia Blumenthal, *Saving in Postwar Japan*
36. Peter Frost, *The Bakumatsu Currency Crisis*
37. Stephen C. Lockwood, *Augustine Heard and Company, 1858–1862*
38. Robert R. Campbell, *James Duncan Campbell: A Memoir by His Son*
39. Jerome Alan Cohen, ed., *The Dynamics of China's Foreign Relations*
40. V. V. Vishnyakova-Akimova, *Two Years in Revolutionary China, 1925–1927*, tr. Steven I. Levine
41. Meron Medzini, *French Policy in Japan during the Closing Years of the Tokugawa Regime*
42. *The Cultural Revolution in the Provinces*
43. Sidney A. Forsythe, *An American Missionary Community in China, 1895–1905*
44. Benjamin I. Schwartz, ed., *Reflections on the May Fourth Movement: A Symposium*
45. Ching Young Choe, *The Rule of the Taewŏn'gun, 1864–1873: Restoration in Yi Korea*

46. W. P. J. Hall, *A Bibliographical Guide to Japanese Research on the Chinese Economy, 1958–1970*
47. Jack J. Gerson, *Horatio Nelson Lay and Sino-British Relations, 1854–1864*
48. Paul Richard Bohr, *Famine and the Missionary: Timothy Richard as Relief Administrator and Advocate of National Reform*
49. Endymion Wilkinson, *The History of Imperial China: A Research Guide*
50. Britten Dean, *China and Great Britain: The Diplomacy of Commerical Relations, 1860–1864*
51. Ellsworth C. Carlson, *The Foochow Missionaries, 1847–1880*
52. Yeh-chien Wang, *An Estimate of the Land-Tax Collection in China, 1753 and 1908*
53. Richard M. Pfeffer, *Understanding Business Contracts in China, 1949–1963*
54. Han-sheng Chuan and Richard Kraus, *Mid-Ch'ing Rice Markets and Trade, An Essay in Price History*
55. Ranbir Vohra, *Lao She and the Chinese Revolution*
56. Liang-lin Hsiao, *China's Foreign Trade Statistics, 1864–1949*
57. Lee-hsia Hsu Ting, *Government Control of the Press in Modern China, 1900–1949*
58. Edward W. Wagner, *The Literati Purges: Political Conflict in Early Yi Korea*
59. Joungwon A. Kim, *Divided Korea: The Politics of Development, 1945–1972*
60. Noriko Kamachi, John K. Fairbank, and Chūzō Ichiko, *Japanese Studies of Modern China Since 1953: A Bibliographical Guide to Historical and Social-Science Research on the Nineteenth and Twentieth Centuries, Supplementary Volume for 1953–1969*
61. Donald A. Gibbs and Yun-chen Li, *A Bibliography of Studies and Translations of Modern Chinese Literature, 1918–1942*
62. Robert H. Silin, *Leadership and Values: The Organization of Large-Scale Taiwanese Enterprises*
63. David Pong, *A Critical Guide to the Kwangtung Provincial Archives Deposited at the Public Record Office of London*
64. Fred W. Drake, *China Charts the World: Hsu Chi-yü and His Geography of 1848*
65. William A. Brown and Urgunge Onon, translators and annotators, *History of the Mongolian People's Republic*
66. Edward L. Farmer, *Early Ming Government: The Evolution of Dual Capitals*
67. Ralph C. Croizier, *Koxinga and Chinese Nationalism: History, Myth, and the Hero*
68. William J. Tyler, tr., *The Psychological World of Natsumi Sōseki*, by Doi Takeo

69. Eric Widmer, *The Russian Ecclesiastical Mission in Peking during the Eighteenth Century*
70. Charlton M. Lewis, *Prologue to the Chinese Revolution: The Transformation of Ideas and Institutions in Hunan Province, 1891–1907*
71. Preston Torbert, *The Ch'ing Imperial Household Department: A Study of its Organization and Principal Functions, 1662–1796*
72. Paul A. Cohen and John E. Schrecker, eds., *Reform in Nineteenth-Century China*
73. Jon Sigurdson, *Rural Industrialization in China*
74. Kang Chao, *The Development of Cotton Textile Production in China*
75. Valentin Rabe, *The Home Base of American China Missions, 1880–1920*
76. Sarasin Viraphol, *Tribute and Profit: Sino-Siamese Trade, 1652–1853*
77. Ch'i-ch'ing Hsiao, *The Military Establishment of the Yuan Dynasty*
78. Meishi Tsai, *Contemporary Chinese Novels and Short Stories, 1949–1974: An Annotated Bibliography*
79. Wellington K. K. Chan, *Merchants, Mandarins, and Modern Enterprise in Late Ch'ing China*
80. Endymion Wilkinson, *Landlord and Labor in Late Imperial China: Case Studies from Shandong by Jing Su and Luo Lun*
81. Barry Keenan, *The Dewey Experiment in China: Educational Reform and Political Power in the Early Republic*
82. George A. Hayden, *Crime and Punishment in Medieval Chinese Drama: Three Judge Pao Plays*
83. Sang-Chul Suh, *Growth and Structural Changes in the Korean Economy, 1910–1940*
84. J. W. Dower, *Empire and Aftermath: Yoshida Shigeru and the Japanese Experience, 1878–1954*
85. Martin Collcutt, *Five Mountains: The Rinzai Zen Monastic Institution in Medieval Japan*

STUDIES IN THE MODERNIZATION OF THE REPUBLIC OF KOREA: 1945–1975

86. Kwang Suk Kim and Michael Roemer, *Growth and Structural Transformation*
87. Anne O. Krueger, *The Developmental Role of the Foreign Sector and Aid*
88. Edwin S. Mills and Byung-Nak Song, *Urbanization and Urban Problems*
89. Sung Hwan Ban, Pal Yong Moon, and Dwight H. Perkins, *Rural Development*

90. Noel F. McGinn, Donald R. Snodgrass, Yung Bong Kim, Shin-Bok Kim, and Quee-Young Kim, *Education and Development in Korea*
91. Leroy P. Jones and Il SaKong, *Government, Business, and Entrepreneurship in Economic Development: The Korean Case*
92. Edward S. Mason, Dwight H. Perkins, Kwang Suk Kim, David C. Cole, Mahn Je Kim, et al., *The Economic and Social Modernization of the Republic of Korea*
93. Robert Repetto and Kwon Tae Hwan, *Economic Development, Population Policy, and the Demographic Transition in the Republic of Korea*
94. Parks M. Coble, *The Shanghai Capitalists and the Nationalist Government, 1927–1937*
95. Noriko Kamachi, *Reform in China: Huang Tsun-hsien and the Japanese Model*
96. Richard Wich, *Sino-Soviet Crisis Politics: A Study of Political Change and Communication*